THE ANT AND THE FERRARI

THE ANT AND THE FERRARI

LIFTING THE HOOD ON TRUTH, SOCIETY AND THE UNIVERSE

DR KERRY SPACKMAN

*I'd like to thank Sabine Tyrvainen for her
tremendous support and countless suggestions
which have made such an improvement.*

HarperCollins*Publishers*

First published in 2012
by HarperCollins*Publishers (New Zealand) Limited*
PO Box 1, Shortland Street, Auckland 1140

Copyright © Kerry Spackman 2012

HarperCollins*Publishers*
31 View Road, Glenfield, Auckland 0627, New Zealand
Level 13, 201 Elizabeth Street, Sydney, NSW 2000, Australia
A 53, Sector 57, Noida, UP, India
77–85 Fulham Palace Road, London W6 8JB, United Kingdom
2 Bloor Street East, 20th floor, Toronto, Ontario M4W 1A8, Canada
10 East 53rd Street, New York, NY 10022, USA

National Library of New Zealand Cataloguing-in-Publication Data

Spackman, Kerry.
The ant and the Ferrari : lifting the hood on truth, society and the Universe / Kerry Spackman.
ISBN 978-1-86950-959-0
1. Teleology. 2. Causation. 3. Truth. 4. Ethics.
I. Title.
124—dc 22

ISBN: 978 1 86950 959 0

Publisher: Alison Brook
Cover and internals design by Athena Sommerfeld
Typesetting by Springfield West
All images courtesy of shutterstock.com except for images from Science Library (pages 11, 31,
110); NASA (pages 28, 74, 139); Wikipedia (pages 32, 81, 128); freemovieposters.net (page 49);
King's College, Cambridge (page 80); and YouTube (page 167).
Colour reproduction by Graphic Print Group, South Australia
Printed by RR Donnelley, China, on 128gsm Matt Art

CONTENTS

CHAPTER I
THE ANT AND THE FERRARI

I want you to imagine a really close-up shot of an ant walking on a shiny red surface. Because the zoom is so tight the ant's tiny body fills your entire view. At this range you can even see its minute antennae waving around. You notice its six little legs struggling for grip because the red surface on which it walks is so highly polished.

Now allow the view in your mind to slowly zoom out. As the image widens notice how the ant becomes smaller and smaller while the vast red surface on which it's walking begins to fill your vision. Before long the ant is just a black dot on a desert of shiny red emptiness — stretching without end. On this scale the ant has become insignificant. And as you let your field of view widen even further you notice the shiny flat surface actually starts to curve. It begins to form a shape you recognize — the hood of a Ferrari.

This image reminds us there are many things the ant will never understand no matter how hard it tries. Its view of the world is so limited it has no idea what lies beneath the bonnet or even that there is a world existing under the hood at all. Its senses simply aren't powerful enough to see the overall picture. But even if we

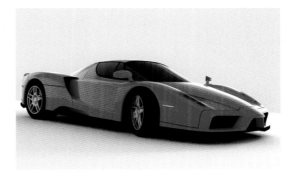

could somehow show the ant all the complex electronics and hydraulics hidden under its feet, it would never understand how they work because an insect's brain is just too small to comprehend all that information.

It's obvious from this story the ant understands only a tiny fraction of the Universe and how it works. But are our brains equally limited? Maybe we'll never know more than 1 per cent of how the Universe works either. After all, it would be a surprising coincidence if our brains were exactly the right size to understand everything, particularly when it's clear every other animal only knows a fraction of what's going on around it. Maybe we are also just like ants crawling across the bonnet of the Universe — blissfully unaware of how the vast majority of the Universe really works and missing out on the most interesting aspects of our short lives.

This book is partly about 'lifting the hood' to reveal all the interesting things normally hidden underneath so you can have a richer and more fulfilling life. Instead of just being carried along for the 'ride' you'll be able to understand the 'how' and the 'why' of the journey.

More importantly, this book is about *Truth* — about what is *actually* under the hood rather than what our common sense tells us to believe is under there. For we'll soon find out our common sense and unaided logic are little more effective than the ant's vision. Instead, we'll need some metaphysical tools to help us see further and in more detail. Doing so will allow us to discover the purpose and meaning of life.

Now if you're like the thousands of people I've spoken to over the years, you'll probably raise some objections as soon as I mention the word 'Truth'. The most common objections usually go along the following lines:

- There's no such thing as 'absolute' Truth. All Truth is relative. What's true for you isn't necessarily true for me.
- What anyone calls Truth is just a human approximation to Reality.
- Science is just another 'belief system' and is no more valid than any other belief system.
- What makes you think you've got a more valid or special way of finding Truth than I have?
- The only way we can know Absolute Truth is by revelation from a higher being. Humans are just too limited to ever discover Absolute Truth by themselves.
- There are many paths to Truth and it is arrogant to think anyone can say which way is better.

I'm not sure if any of those ideas had any resonance with you, but they're all extremely serious issues and so we'll treat each of them with the respect they deserve. As we do, we'll use them to answer some of the biggest questions that have puzzled mankind since the beginning of time:

- What is the origin of the Universe? Was a creator involved or can the Big Bang explain everything?
- Is there life after death? Can we ever know?
- How should we organize society and what are our greatest political and economic threats to peace and happiness? Has capitalism had its day?
- If every event has a cause then what caused God or the Big Bang?
- Can you really have 'free will' and a 'destiny' at the same time?
- Is Evolution really able to produce something as complex as a human or is there still something missing?
- Is there any basis for ethics or is it everyone for themselves?

PLAY LIFE BY THE WRONG RULES

The reason why these questions are so important is because our beliefs influence every aspect of our life. If our beliefs don't match Reality, it would be like playing the game of life by the wrong rules. For sure you'd be upset if you were playing blackjack and had bet your entire life savings on what you thought was a winning hand, only for the casino owner to tell you he'd 'changed the rules' and now 18 was a winning hand instead of 21.

It's the same with life. If our beliefs don't match Reality — whatever 'Reality' is — we're likely to end up with a losing hand. Unfortunately, we often acquire our beliefs when we are young and least equipped to audit their validity. But once inside us, even demonstrably false beliefs are remarkably immune to contrary evidence because of the way our brains work. As we'll see in Chapter 12, The

Psychology of Belief, our beliefs distort Reality so that any conflicting information is twisted to match our beliefs and keep them intact. It's as if we wear 'blue mental glasses' for 'blue beliefs' and 'red mental glasses' for 'red beliefs'. That's part of the reason why our own beliefs always seem so sensible to us while everyone else's seem so silly.

As I mentioned before, one of the purposes of this book is to develop tools which will allow us to take off our tinted glasses and see things unhindered by our human limitations.

> 'The door of the prejudiced mind opens outward so that the only result of the pressure of facts upon it is to close it more tightly.' — Ogden Nash

A PERSONAL CHALLENGE

By now you might be getting a little sceptical as to whether this book can actually live up to its promise. After all, bookstores are packed with books claiming to make you thinner, richer, happier, wealthier and lead you to the path of true enlightenment and most of them don't live up to their wild claims. Why should this book be any different?

The best answer I can give is that after finishing the first draft of this book, I sent out what I called my 'Editor's Draft' to as wide a range of people around the world as I could find. I wanted to get feedback from surfer hippies, Buddhist monks, mums looking after children, scientists, business people, the unemployed, philosophers, writers, teenagers and people who wouldn't even normally read this type of book. The only thing I asked in return was that they filled out an anonymous survey on what they liked and didn't like so I could improve the book. To make it easy for everyone, I set up a private little website where they could leave their comments about any aspect of the book. That way they could be totally open and as critical as they liked because I'd never know who'd written each comment.

Of course there were a number of surprises as to what was popular and what wasn't, but one thing stood out that I wasn't expecting. Every single person who read the book — 100 per cent of readers — said this book had changed some aspect of their 'belief' or made them think about something in a different way. Whether it was about Religion, Science, Ethics, Beauty, Politics, Art, Philosophy or how we structure and order the society we live in, something changed in their mind. Maybe I was just lucky with the people who received my book.

'One of the most important things a person can do in their life is to take in new information and change the way they think.' — Anon

I sincerely hope you will have a similar experience and find ideas in this book which challenge you to view things in a different way. Having said that, I don't want to sound arrogant and I certainly don't claim to have 'all the answers'. I simply want to ask the 'right questions' and hopefully provide some reliable 'tools' you can trust. Tools you can use to test your own beliefs in ways you might not have previously done. And that seems to be the key. In everyday life the correct tools allow us to do far more than we can in our natural unaided state. A shovel allows us to move dirt more efficiently than we can with our bare hands and a telescope allows us to see further than with our naked eye.

In the same way, I hope the tools in this book will open up new vistas and allow you to enjoy a richer, wider view of life. I hope that you will be like the man in the Flammarion engraving in the picture above, still firmly in our natural world where there are stars, trees, rivers and cows, but somehow you manage to poke your head out into the hidden unknown world and see how it all works.

'What I see in Nature is a magnificent structure that we can comprehend only imperfectly, and that must fill a thinking person with a feeling of humility. This is a genuinely religious feeling that has nothing to do with mysticism.' — Albert Einstein

CHAPTER 2
YOUR MOST PRECIOUS GIRL

Before we get started, I'd like to give an even stronger reason why I think our 'beliefs' are the most important things in life. To do this, I want you to think of the one young girl who is more precious to you than any other girl in the world. If you have a daughter, it will obviously be her, but if not she could be someone close, like a niece or a friend's child. Ideally, she'll be aged around seven years old, but she could be a few years either side of that. If you don't know anyone who is this age at the moment, go back in time and think of someone from your past. It doesn't matter who this girl is as long as you care deeply about her.

So please, take a moment to select your girl and then bring her personality and image vividly to life in your mind. Doing this will be crucial if the next paragraph is to be more than mere words printed on a page.

Now I want you to imagine you and your special girl are alone in a quiet park. She's climbing up and down on a jungle gym on the far side of the park while you sit watching her from a distance on a bench in a cool, shady corner. You hear her excited squeals of joy as she conquers one challenge after another and you see the obvious delight on her face. It's a lovely summer's day and the sweet scent of freshly mown grass reminds you of your own childhood all those years ago. So you close your eyes and before long you're back in those innocent days of your youth when there was nothing to worry about. A wonderful warm, peaceful feeling sweeps over you.

A moment later you open your eyes and you're shocked to see a man placing his rough hand over the mouth of your precious child as he begins to drag her towards the bushes. A sharp pain instantly shoots through your stomach and time switches into slow motion. Nothing seems real. For a split second you are frozen ... stuck to the bench unable to move as you grasp the reality of the situation ... until finally you scream out at the top of your lungs ... running towards your precious little girl ... every muscle in your body ... every fibre straining ... faster and faster ... yelling ... running ... yelling. Fortunately, your actions scare away the would-be rapist and you reach your precious little girl. You hold her ever so tightly against your pounding chest.

She's safe.

This disturbing story illustrates just how much effort you'd be willing to expend to protect your precious child — in fact any child — who might be harmed like this. Without question you'd do absolutely everything in your power to ensure her safety. This brings us to the central question of this chapter: *What would you do if instead of there being a **person** who was about to harm your precious child, there was a **belief** which would cause her exactly the same amount of damage?*

Surely, once again you'd do everything in your power to prevent such a dangerous idea from spreading. After all, if the consequences were exactly the same, it shouldn't matter if it was a belief or a person who was about to harm your child.

In order to show this isn't just some hypothetical argument but is a question which has genuine consequences, let me give you a real-life example of what I'm talking about. I'll choose an example that's related to our story of the little girl above, but even though I do, it's important to realize this book is not about the horrors of sexual abuse but rather is about beliefs, truth and reality. I've simply chosen the next example because it ties in with the story of our little girl and illustrates the importance of beliefs and their consequences.

THE CONSEQUENCES OF FALSE BELIEFS

In Africa there are many men who genuinely believe the cure for AIDS is to have sex with a young virgin, and the younger the girl is the more powerful this cure is supposed to be. Clearly, this belief is utterly false, and instead of curing AIDS, it has the exact opposite effect because it spreads the disease even further. What makes this belief even more dangerous than a 'child molester' is that a 'false idea' like this can spread throughout society causing catastrophic damage on a truly massive scale. Indeed, according to some estimates, more than 20,000 girls in Africa are raped because of this belief every year.

So let me come back and ask you again: *If you would strain every muscle and fibre in your body to protect your precious little girl from some would-be attacker, then wouldn't you put the same effort into defeating a false belief like this one about AIDS if you were able to do so?* Of course you would, because by protecting the Truth like this you'd save thousands of real, living, breathing, innocent girls.

This concept of beliefs having far-reaching consequences covers a vast range of topics affecting every aspect of our lives, from Religion to Medicine and Politics to name but a few. For example, if a person believes the cure for cancer is chemotherapy then they're likely to have a different outcome to another person who believes the cure is magic crystals. Likewise in Politics: a country which embraces capitalism is likely to have a different standard of living and well-being to a country which adopts communism. With those two examples it's easy to observe the outcome, but with other beliefs the true consequences are often related in a second- or third-hand manner which makes them much more difficult to spot.

MORE POWERFUL THAN NUCLEAR WEAPONS

If you think about it, beliefs are even more powerful than nuclear weapons because they shape the actions of everyone who lives on our planet. You only have to think of the terrible events of September 11 2001 when almost 3000 people were killed by Al-Qaeda extremists because they genuinely thought this was the will of their God. These people — just like your brother or sister or your best friend — were killed by a belief.

These examples bring home the point that beliefs don't just operate inside some 'dry academic vacuum' where they only interact with other beliefs. Instead, they influence every aspect of our daily lives. Of course not all beliefs are as catastrophic as these and some only affect the small details of our lives while other beliefs are genuinely helpful. But the point is, whenever we tolerate any belief which doesn't match Reality — how the Universe actually works — we run a risk of causing damage to society. And yet while we go to such enormous lengths to protect nuclear weapons by shielding them in massive concrete bunkers, we let society play fast and loose with Truth.

We seem to think it's OK for people to believe anything they like regardless of whether or not their beliefs actually match Reality. Somehow in our modern world we've confused 'freedom of speech' with the idea that any belief goes. So instead of confronting false beliefs and examining their foundations, we end up pussyfooting around people who believe provably false ideas because we don't want to upset them. It's no wonder society is in such a mess. Instead of confronting a radical religious belief head on, we spend billions of dollars on airport security and an endless 'war on terrorism' while the underlying belief is left intact. Not surprisingly, as soon as we stamp out one terrorist threat another rises up because the core has not been dealt with.

CHAPTER 3
TOOLS FOR THE JOURNEY

You wouldn't attempt to climb Mt Everest without first ensuring you have all the tools you need to achieve the summit: oxygen tanks, sufficient food, ice pick, warm clothes and so on. Neither should we begin our search for Truth unless we know we have sufficient 'equipment' to complete the task. Therefore the first thing we need to do is consider our own natural resources to see if they are sufficient.

Let's start by seeing how reliable our brain is. If you look at the two images below you will see the left image has five dots which appear to 'pop out' of the page (convex shape) and one dot which appears to 'pop in' to the page (concave). But on the right image you will see the exact opposite. Five dots appear to 'pop in' and one 'pops out'.

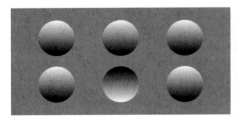 

Now the interesting thing about these two images is that they're exactly the same — except for one thing. I've simply inverted the image on the right. You can check this by turning the book upside down and then you'll see that the dots 'pop' in the opposite direction for each picture. The question is: Why does our brain do this and what does it mean?

The answer hinges around your brain's limited processing power and all the millions of things it needs to do at the same time. Every millisecond of the day your brain is incredibly busy doing things such as processing all the images in your visual field, controlling your muscles to keep you balanced, and carrying on that conversation with your best friend. Each of those individual tasks is beyond the most powerful supercomputers we've ever made. The only way our brain can accomplish all these tasks at once is to take lots of short cuts. Otherwise we'd have to carry around such an enormous brain we wouldn't be able to eat enough food to keep it supplied with sufficient energy.

The particular short cut your brain uses in the 'dots' example is to always assume the sun (or a light bulb) is above you. This is a very good assumption because throughout your entire life the sun has always been above you. And if the sun is above you, then any object that 'sticks out' will be bright on the top and dark on the bottom (because of a shadow) as the diagram 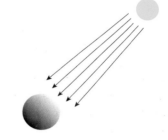 shows. Conversely, any concave dot that pops in will be dark on the top and shiny on the bottom. If you now take a look back at the pictures of the dots you'll see the ones which pop out are indeed shiny on the top and dark on the bottom.

The reason your brain uses this short cut is because it's much easier to work out if something is bright on the top or on the bottom from a distance, than it is to calculate the actual curvature of an object. It's a bit like trying to work out whether a car is a Ferrari or a Lamborghini from a kilometre away. From that distance it's very hard to do because they're both low-slung sports cars with similar shapes. But *if* Ferraris were always red and Lamborghinis were always yellow, it would be a piece of cake. You'd just use that colour 'short cut' without having to look at all the fine differences between them.

This brings us back to our journey to search for Truth. We wouldn't have to worry about our brains' short cuts if those short cuts always gave us the perfect answer and accurately reflected what was actually real. Unfortunately, these short cuts often lead us astray without us even knowing it. Take a look at the checkerboard over the page constructed out of a mosaic of dark and light squares. For convenience I've marked one of the dark squares with the letter A and one of the light squares with the letter B. Your job is simply to decide how much darker the square with the letter A on it is than the square with the letter B on it.

The best way to think about this is to try to guess how much more ink was used to print the square with the A than was used to print the square with the B on it. If you think twice as much ink was used, you would say the A is twice as dark as the B. If you think it used 30 per cent more ink, you would say it's 30 per cent darker. Have a go.

Edward H. Adelson

It turns out the two squares A and B are *exactly* the same! If you don't believe me get a piece of paper and place it on top of the picture to cover it up. Then punch two small holes in the paper so you can only see the A and the B squares and nothing else. You may be surprised at what you see.

The reason why I've shown you these two optical illusions is to highlight important issues we need to be aware of as we search for Truth:

1. Your brain does many things without you knowing about it.
2. Sometimes it gets the wrong answer even though you're absolutely and thoroughly convinced you're right.

And being wrong when you're absolutely convinced you're right is the last thing you want when you're on a search for Truth. Unfortunately, things get even worse than this. Not only do our senses mislead us but our logic often does an even worse job without us realizing it. Like everyone else, you probably think you don't make errors of logic or reason, but when we come back to this in a later chapter you might be surprised. So, given all our natural limitations what are we to do? The answer is we need tools to help us.

THE COMPASS OF TRUTH

One tool I've found particularly helpful is what I call 'The Compass of Truth'.

In Chapters 7 and 8 I'll show you how to assemble your own Compass of Truth, piece by piece, out of rock-solid logic. Because you'll assemble each part yourself, you'll know exactly what's inside it and how each component works. That knowledge will give you confidence in its ability to guide you through life's decisions and help answer some of the most important issues of today. You'll also find it particularly useful when debating issues with other people and it will help you to answer those tricky questions at the start of this book.

> 'I just want to start a conversation so we shall all feel at home with one another like friends.' — Socrates, as he commenced laying the foundations of Western philosophy

CHAPTER 4
THE *MYTHBUSTERS'* ORIGIN
OF THE UNIVERSE

WHAT ARE WE DOING HERE?

Humanity has always wanted to know how we got here and what our place in the Universe is. Are we just a random accident or do we have a purpose — perhaps even a divine purpose spelt out by the stars? Because our origin defines who we are, it's no wonder that most religions have an account of creation followed by a description of our status in the Universe.

In the beginning God created the heavens and the earth ... Then God said, 'Let there be light'; and there was light. God called the light day and the darkness He called night. And there was evening and there was morning which was the first day.

And God said, 'Let the waters under the heaven be gathered together unto one place, and let the dry land appear.' And it was so. And God called the dry land Earth; and the gathering together of the waters he called Seas: and God saw that it was good. ... And the evening and the morning were the third day.

And God made two great lights; the greater light to rule the day, and the lesser light to rule the night: he made the stars also. And God set them in the firmament of the heaven to give light upon the earth, and to rule over the day and over the night, and to divide the light from the darkness: and God saw that it was good. And the evening and the morning were the fourth day. — Genesis 1:1

Now any description of creation, like the one from the Bible above, must fall into one of three categories:

1. It must be *true*.
2. It must be *false*.
3. It must be *neither true nor false*.

There are no other possibilities.

The third option of being *neither* is where you'd put all interpretations of Genesis that say it's an allegory or myth: a story that's meant to illustrate a point or moral and therefore shouldn't be taken literally. If you place it in this category then you are saying Genesis makes no specific claims about what actually happened.

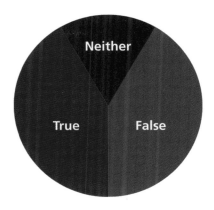

The question then arises as to whether or not we can determine which category the Genesis account of creation actually falls into. Is it *true, false* or *neither*?

This question deeply divides Americans, as indicated by a recent Gallup poll which revealed that:

1. 53 per cent believe God created the heavens, the Earth and mankind in six literal days — exactly as described in Genesis above.
2. 31 per cent think humans evolved over many millions of years, but that God directed or influenced the evolutionary process. In other words, Evolution wasn't random or blind; there was an overseeing hand of God.
3. 12 per cent of Americans think humans evolved from simple cells over a period of millions of years and that God played no part in the origin of humankind.
4. 4 per cent think some other mechanism was involved, such as reincarnation, or they didn't know how we got here at all.

Obviously these four beliefs can't all be right at the same time, and that means millions of people must be going about their daily lives busy making their plans and basing their decisions on an error like our gambler playing by the wrong rules.

The reason why knowing our origin is so important is because this knowledge fundamentally determines our value, meaning and purpose in life. If we were created by a God who cares about us and has a personal plan for us, that gives us one type of status or purpose. But if we evolved by Random Chance without any plan, we'll 'feel' entirely differently about the meaning of our life. That's why this issue has spawned so many arguments.

We'll need to revisit this connection between 'design' and the 'value of our life' later on in Chapter 18 (Purpose and Meaning), because there's a very surprising twist in the tail. The connection between our purpose and any design is not what most people think. But let's put this aside for a moment and carry on.

Trying to understand our origin and status is the oldest question humankind has asked. It goes back to the dawn of civilization. Even today the search for an answer to this most important of all questions continues. According to the internet search engine Ask.com, out of all the billions of internet searches in 2010, the number one and number two most frequently asked questions were:

1. What is the meaning of life?
2. Is there a God?

So let's begin our search for Truth by seeing if we can prove where the Genesis account of creation should fit. Is it true, false or neither? After all, any book claiming to be about Truth should at least be able to answer that question.

Unfortunately, if you look at the four options in the Gallup poll, you'll immediately see we face a major problem finding an answer that's going to satisfy everyone. Option 3 seems to be framed in 'scientific language' whereas option 1 is based on 'religious belief'. A religious person could quite easily object to any scientific argument by saying something along the lines of: 'Well, Science is just another belief system and I choose my Christian belief system over any scientific one.' Or, 'I have faith in my religion and this is enough for me. Faith supersedes Science. Even if some things don't seem to agree with my belief, I know my religion works for me.' Conversely, a scientist could equally say something like: 'The only things I believe in are those based on hard scientific evidence and "faith" is simply the absence of evidence.'

Given these two different starting points it's hard to see how any approach is going to satisfy everyone. Scientists will rely on their 'empirical evidence' and believers in their 'faith' or 'personal experience'. This situation is a bit like two people arguing over who's the best sportsperson. One person insists the best way for them to decide is by playing each other in a game of tennis while the other person thinks the best method is a golf challenge. Unless they can agree on what is a valid test they'll never make any progress.

In the same way, believers and scientists often appear to be playing different 'games' and to rely on different types of evidence. This is a serious challenge, which is why we'll need to develop our Compass of Truth in Chapters 7 and 8 to steer us through this tricky issue.

MYTHBUSTERS

You've probably seen the very popular TV series *MythBusters*, where Jamie and Adam test popular myths to decide if they are 'Confirmed, Busted or Plausible'. It's a fun and easy format to follow, so let's use their approach to begin our investigation. After all, at least three of the options in the Gallup poll must be myths. Let's see which are Confirmed, Busted or Plausible.

In typical *MythBusters'* fashion we'll start by blowing up a car. (They love their explosions!)

Like Jamie and Adam, we'll replay the car explosion in ultra-slow motion. When we do we see thousands of fragments flying in all directions: a bit of a door going this way, a wheel going that way. But rather than just observing the overall carnage, we're going to analyse what happens to each of these pieces, one at a time.

The best way to do this is to pause the video and follow each fragment as we advance the video frame by frame. Doing this allows us to calculate the speed and direction of motion of every individual piece using the following technique:

1. We determine the speed of each fragment by seeing how far it moves between successive frames. For example, a door might move 3 metres between two frames while a wheel might move 6 metres. If that happens the wheel must be going twice as fast because it has travelled twice as far as the door in the same time.
2. We determine the direction each object is moving in by drawing a line that joins its position as it moves from one frame to the next.

When we overlay these speeds and directions directly onto the video footage we notice something very interesting. No matter which frame we look at, the fragments furthest away from the car are going the fastest and the items closest to where the explosion began are going the slowest. But it's even more precise than that. An object twice as far away is going twice as fast. An object 3.24 times as far away is going exactly 3.24 times as fast.

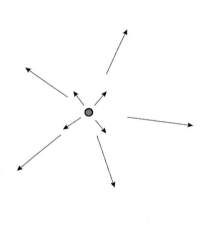

We also notice that each particle is moving in a direction pointing directly away from one single point. And what's so special about that point? It's exactly where the car used to be. This is shown schematically in the top diagram on the right.

If we rewind our video and play it back in reverse we find all the fragments fly back together into a single place, at exactly the same time, to form a car. It's like going back in time. But if any of the directions are misaligned, the objects won't all meet back at the same place. Similarly, if some

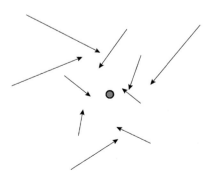

of the speeds are too fast those objects will arrive too early and carry on past the central point before the other fragments arrive. Again, we won't get everything back together in one nice car.

The only way we can get everything back together in one piece is if:

- All objects move directly towards a
  single point. Their angles must be
  perfectly aligned.
- The speed of every object is
  perfectly matched with how far
  away it is from the central point.

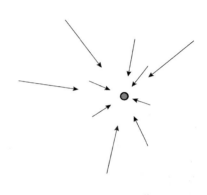

Items twice as far away must be going
twice as fast. Items 2.41 times as far
away must be going 2.41 times as fast.
The match of speed and distance must be
exact — otherwise they won't arrive at the
same time.

Now supposing we hadn't seen the whole video of our car explosion, but were only given a single frame recorded halfway through the explosion. If all the speeds and directions of every fragment were marked on that frame, and they all perfectly matched our requirement of speed and direction, we could reasonably conclude all the fragments must have been in the same place at the same time in the past. That is, they came from a single explosion.

BLOWING UP THE UNIVERSE

This brings us to a most surprising and unexpected discovery.

In the 1930s, astronomer Edwin Hubble decided to measure the speed of the stars and galaxies using the same technology the police use when they check your car's speed with a radar gun. Hubble was gobsmacked to discover exactly the same pattern with the stars as we've just seen with our exploding car: stars that were twice as far away were going twice as fast and galaxies that were 127 times as far away were going 127 times as fast. This was a total shock to him, because at the time everyone expected all the stars to be fixed nicely in their own place in a static Universe.

Since then, thousands of people have checked literally billions of stars and galaxies using a variety of different speed-measuring techniques — and they always find the match of distance and speed is exact. We can be as confident of the stars' speed as the police can be of your car's speed. If we don't doubt the police radar guns then we shouldn't doubt the speed of the stars.

What was even more surprising was to find that all the stars were moving away from a single point, just as the fragments in our car explosion were. They're all perfectly aligned. This can't possibly be a coincidence. The only sensible conclusion we can come to is that at some time in the past everything in the Universe was squashed together in one place.

We can easily calculate how long ago that was, by dividing the distances to each star by their speeds. It's the same technique we use when calculating how long a journey is going to take. We divide the distance of the journey by how fast we're going to drive. If it's 300 km and we're going to drive at an average of 60 km/h then it will take us 300/60 = 5 hours. When we do this calculation for the stars we find the entire Universe — every star and every galaxy — was squashed together into a single point 13.7 billion years ago.

It's looking suspiciously like some sort of big explosion took place.

THE HEAT OF THE EXPLOSION

But there's something else we know about explosions. When things explode, we don't just have bits and pieces flying everywhere, we also observe that explosions give off a tremendous amount of heat. Gradually, this heat fades with time — and the longer we wait the cooler things will be. This 'cooling down with time' is something we're all familiar with. It's why we wait for a cup of coffee to cool down before we drink it. Which raises a question: If the entire Universe did blow apart in a big explosion (and we can't yet be sure it did), is it possible to measure the heat left over from that explosion?

Unfortunately, we've just worked out the Universe exploded 13.7 billion years ago. That's an awfully long time ago and so there's not likely to be much heat left over today. Or is there? We might just have a chance, because another thing we notice in everyday life is that the bigger and the hotter something is the longer it takes to cool down. Well, the Universe is certainly big. In fact it's the biggest thing possible. And any explosion big enough to blow trillions of stars trillions of miles apart must have been truly gigantic. Maybe, there might just be some flicker of heat left over today.

Back in 1948 physicist George Gamow decided to find out. He first calculated how big the explosion would have to be if it was going to blow all those trillions of stars apart at the speed we see them going at today. It's the sort of calculation you need to do when you work out how much dynamite it will take to blow up a building.

Gamow did his calculations and worked out how much heat would be left over today if it cooled down for 13.7 billion years. He came up with a very chilly −270 degrees Celsius (or −460 Fahrenheit). That's only three degrees above the coldest

possible temperature called Absolute Zero. This was rather depressing because he couldn't possibly hope to find such a tiny amount of temperature in outer space when the Earth on which he was standing was +20°C. That's a full 290°C hotter than the flicker of temperature he was looking for. It would be like trying to listen to someone whispering 100 miles away when you're at a rock concert surrounded by ear-shattering music.

So everyone forgot about Gamow and his tiny −270 degrees of 'leftover' temperature.

But 17 years later in 1965, two scientists who'd never heard about this 'leftover' heat became rather upset when they turned on their new radio telescope to look at the stars. Everywhere they looked, they noticed their images were blurred by a faint heat glow of −270°C. At first they thought it was a problem with their receiver and so they tried everything they could to get rid of this extra heat. Gradually, it dawned on them that this temperature must be real and filling the entire Universe.

But why? Where had it come from?

They had no idea until someone told them about Gamow's old 1948 calculation. Suddenly, it all made sense. Since then thousands of people have independently measured this leftover temperature using a wide range of different instruments. Some of these measurements were taken from Earth, some from space. They all give the same answer. We now know this leftover 'background temperature' is precisely −270.425°C. This couldn't possibly be a coincidence and it matches the size of the explosion and the age of the Universe to an incredible accuracy.

Again, we can be very sure of these measurements because we routinely use 'thermal imaging' in our everyday lives here on Earth as the thermal image of the house below shows.

But even more telling than the temperature being correct is that it varies by less than 0.00001°C as it fills the entire Universe. This incredible consistency across the vastness of the Universe rules out the heat coming from any other source such as from the stars. If the heat was from stars then it should be hotter where the stars were and colder where there were empty holes in space. But if we take out the heat from all the stars we find a perfectly smooth temperature left over. The only logical conclusion is that it must have come from a single explosion at the centre and start of the Universe and then been stretched smooth as the Universe expanded and cooled.

So we're faced with a set of independent observations that all seem to fit perfectly together. We see the entire Universe flying away from a single point and by tracking the speed of stars and galaxies we see they all appear to have come from the same point at the same time 13.7 billion years ago. And we find, quite by accident, exactly the right amount of heat left over and spread out in just the way we'd expect if it was driven by a colossal explosion.

But not just any explosion; one that was exactly the right size — and at exactly the right time.

There's an even more compelling reason to believe the Universe exploded 13.7 billion years ago. A reason that depends on a little cooking!

BAKING BREAD

Let's consider what happens when we cook a loaf of bread. As the dough bakes, the heat from the oven causes the ingredients to undergo chemical reactions. What happens to the dough depends on how hot the oven is and how long we leave the bread in there. If the oven is too hot we end up with a burnt outer crust and a gooey inside.

Conversely, if we cook the bread too slowly it dries out, becoming a hard solid lump with no crust at all. As Gordon Ramsay will tell you, getting the temperature and timing right is an essential part of being a good cook. And it also works the other way around. A top chef will be able to tell if your oven was too hot or too cold just by looking at the results. It's something we see time and again from the judges on TV programmes like *MasterChef*.

Now this question of 'how hot and how long' sounds exactly like the sort of thing we've just been talking about when we discussed the origin of the Universe. Our measurements tell us the Universe is 13.7 billion years old and we also know how hot it was as it cooled down during its expansion. Could our knowledge of 'how hot and how long' tell us anything more about our origin? About who we are?

Well, of course the best answer would be to have an oven that was just as hot as the Universe was when it first exploded, and to fill that oven with the right ingredients. But we'd need an oven that can cook things at thousands of millions of degrees. Fortunately, just such an oven exists, although it doesn't look anything like the oven in your kitchen.

To have a cook-up at those sorts of temperatures we need an oven that is truly huge. The biggest oven in the world is 27 km in circumference and is buried under Switzerland. It's called the Large Hadron Collider (or LHC) and can heat things to 10 trillion (10,000,000,000,000) degrees Celsius! This device is usually called a particle accelerator but we can just think of it as the 'mother of all ovens'. The phenomenal temperatures that can be produced inside this oven allow us to recreate some of the conditions that occurred at the very start of time.

As we turn up the temperature in this oven so it gets closer and closer to the temperatures at the start of the Universe, we suddenly find truly amazing things begin to happen. And when I say amazing, it might be best to give you a little idea of what I'm talking about. It would be like Gordon Ramsay having a completely empty oven with absolutely nothing inside — not even air — and as he turns up the temperature, suddenly out of absolute nothingness, out of a totally empty vacuum, a loaf of bread suddenly appears. And not just any loaf of bread, but one that was just perfect. A nice firm crust with soft moist bread inside.

We can even catch these moments of things being created out of 'nothingness' on film. At one point in space there's nothing and then, suddenly, something appears.

To understand how an oven can perform such a magic trick we'll need to be familiar with one of Einstein's most famous equations. But don't worry — it's not that complicated after all.

EINSTEIN AND E = MC²

Einstein's reputation for brilliance was well-deserved. He single-handedly discovered many truly remarkable equations which describe our Universe and his equations led to many of the inventions we take for granted today — computers, microwave ovens, GPS navigation and so on.

We're going to look at one of his simplest equations: $E = mc^2$.

This equation simply says that matter (**m**) can be turned into pure Energy (**E**) and vice versa.

The equals sign in this equation is really nothing more than a double-headed arrow ←→ so we could write:

$E = mc^2$ means
$E →\ mc^2$ you can turn Energy into matter
$E ←\ mc^2$ you can turn matter into Energy

The **c** in the equation is just a very big number, namely the speed of light in a vacuum, which is 1,000,000,000 km/h and so c^2 is a huge number: 1,000,000,000,000,000,000!

In our everyday life we naturally think of matter — solid things such as people, rocks and water — as being quite different to Energy, something with absolutely no substance at all. But Einstein showed they're actually different aspects of exactly the same thing. A good analogy is to think of ice and water. Water and ice look and act so completely differently to each other that if you'd never seen one turn into the other, you'd never have guessed they were the same stuff. Ice is hard and brittle. You can carve it and shatter it with a hammer while water is a smooth, flowing liquid. Yet they're the same stuff. When you turn ice into water, or vice versa, there's nothing left over. It just changes seamlessly from one into the other.

In the same way, the solid things with substance we call matter are just energy that's crystallized out into a different form. It's only because of our limited ability to see and understand the world through our five senses, and because of our little brain, that we perceive them as being different. But there are a few special places where we can actually see *energy* being turned directly into *matter*. One of them is inside the LHC.

An atomic bomb is another device that does this.

Atomic bombs are machines that convert a tiny amount of hydrogen into a truly huge amount of pure energy. And according to Einstein the amount of **Energy** you'll get is simply the weight of the hydrogen measured in kilograms (which is called its **mass**) multiplied by c^2 or (1,000,000,000,000,000,000). Because of the size of the **c**, we only need to convert the amount of hydrogen contained inside a child's party balloon to blow an entire city apart.

This same nuclear process of turning matter into energy is also how the sun and all the stars work. Stars get their energy by squashing two atoms of hydrogen into each other with such force that they turn two hydrogen atoms into a single helium atom. Because one helium atom weighs less than two hydrogen atoms there's some mass lost during the squashing process. It's this missing mass that's turned into pure energy that powers the sun. Unfortunately, you need a lot of force to squash two atoms into one atom, which is why it's so hard to do. In the sun this force comes from its huge gravity which squashes things under 400 million kilograms per square centimetre of pressure at temperatures of 6000°C.

If the sun didn't run on these nuclear reactions but instead ran on chemical reactions like those that drive our cars or power dynamite it would have run out of fuel long ago. Indeed a single solar flare on the sun has the same power as 40,000,000,000,000 atomic bombs. And yet our sun has been powering away for millions of years. Our very existence depends on the sun, 149 million kilometres away, turning hydrogen into pure energy — day in and day out — as described by Einstein's equation.

CREATING THINGS OUT OF NOTHING

This brings us to the most interesting part of Einstein's equation. As we've seen, the equals sign in $E = mc^2$ means we can go in either direction. Instead of turning matter into energy we can also create matter out of nothing but pure energy. This is a truly mind-boggling concept. We can take absolutely nothing but pure energy and make solid matter out of it.

Just as the sun takes hydrogen and turns it into energy, so too you can perform the reverse process and produce hydrogen and helium out of nothing but pure energy. By now you've probably guessed that creating matter out of energy is going to be something that happened when the Universe started during the Big Bang and is likely to have important implications for who we are.

What's more interesting to us at the moment is not that you can create matter solely out of energy — but rather what sort of matter you get when you do this.

COSMIC COOKING

When we look into this 'mother of all ovens', we find that the type of matter we cook up out of pure energy is extremely sensitive to how hot and how long we run the oven. It would be like setting the oven in your kitchen to 325°C and finding you get a roast chicken produced out of nothing, but when you set the temperature to 324.5°C you get a chocolate cake produced out of nothing. Just the slightest change in temperature and you get completely different results. This is what we observe in the LHC. We can set the machine at different temperatures and we can record on video the different types of particles and atoms we cook up. We find it's a highly predictable oven.

The phrases 'how hot' and 'how long' sound exactly like the two things we've just been discussing about the Universe. We can use our observations of what we see in the LHC oven to predict what would be produced if the Universe:

- exploded 13.7 billion years ago
- expanded at the rate we measure today where the speed and distance of each star has the perfect match we observe
- exploded with exactly the amount of energy we measure
- cooled down just as we think it did.

When we use these measurements we see that 1/100th of a second after the Big Bang the Universe would have cooled down to 100 billion degrees Celsius. At this temperature, pure energy starts to turn into matter. Three minutes and 46 seconds later, when the Universe had expanded and cooled to 900 million degrees, pure energy was no longer able to turn into matter. Even more impressive than the precise time scales we're talking about is that our measurements in the LHC oven predict exactly how much hydrogen, helium, deuterium and lithium should be produced each second during this first 3 minutes and 46 seconds. And it turns out that the slightest change to how we think the Universe exploded would produce dramatic changes in the ratio of these ingredients. This statement is worth repeating. Despite the Universe being billions of years old, we need to get our calculations of how the Universe expanded during the initial phase accurate to seconds! If you thought cooking a soufflé was tricky, getting these ratios right is vastly more exacting.

So, what happens when we look at the Universe through our telescopes? You guessed it. We find the Universe is made up of precisely the right mixture of hydrogen, helium, lithium and deuterium — every different type of atom, in just the right proportion. So where does that leave us?

Well, it's a bit like collecting evidence from a crime scene. We can never be totally sure of anything, but the more independent evidence we can collect that points in the same direction, the more certain we can be of the conclusion. Let's illustrate this point with an analogy.

Suppose we drive around a corner on a sunny afternoon and find a man slumped over the steering wheel of a smashed-up car. Beside the car is a recently broken lamppost with live wires dangling from it. Our initial reaction would be to guess the driver lost control of the car and crashed into the post. We'd become even more convinced if we also discovered fresh skid marks on the road leading straight towards the post and then jerking sharply to the left before ending directly underneath the still smoking tyres of the smashed car. And if a neighbour comes running up the road and says, 'I just heard this big screech of tyres and then a loud bang from over here', you'd be even more certain of your hypothesis. Then a second person comes cycling around the corner and says, 'I saw that red car race past me and hurtle towards this corner at over 100 km/h. I knew immediately it wasn't going to make it.' And finally a third person comes out from the house immediately behind the broken lamppost and says, 'Wow, I had a narrow escape. I was just standing by that lamppost only a minute ago. If I hadn't gone inside I'd probably be dead by now. Look at that lamppost. All smashed to pieces.'

In the same way we know what happens to everyday objects when we blow them apart, and we're pretty comfortable with our radar guns measuring the speed of our cars, so it's fairly convincing, having measured how the Universe moves, that it also came from an explosion, and that this explosion occurred 13.7 billion years ago. And if we add to that our experience of coffee cooling down and leftover heat and then we combine this with what happens inside the 'mother of all ovens' ... well, we'd be pretty silly to say it's all just a big coincidence. And if that was just the start of a long list of another 120-plus completely independent and totally different observations that led to the same conclusion (which is exactly what we find), then surely any reasonable person would have to conclude it looks like the Universe probably exploded apart from a single point around 13.75 billion years ago. No other natural hypothesis seems to fit the facts.

But this is by no means a 'slam dunk'. There are still a number of valid issues which we'll need to address in later chapters:

- What about the idea that Genesis wasn't a factual account anyway and was only supposed to be an allegory illustrating an important moral point? (Chapter 9: Genesis Revisited)
- Can the Big Bang tally with the Bible's description in Genesis of how God created the Universe? In other words, is there a possibility they can still both be true? (Chapter 9)
- Even *if* the Universe did start with a Big Bang, who or what caused the Big Bang in the first place? (Chapter 14: What Caused it All?)
- What about the viewpoint that says regardless of how compelling the Science is I know my God is real because I have a personal and direct relationship with him every day? (Chapter 12: The Psychology of Belief)

CHAPTER 5
IT'S ONLY A THEORY

After reading the last chapter, one of my Christian friends called me and said, 'But of course the Big Bang's *only a theory.*' This is a commonly held objection and so I felt compelled to write the following very short chapter in case you've come across it as well. The problem arises because Science has got itself into a rather tricky situation owing to the unfortunate way we use language.

THE TRAIL OF SPILT MILK

The difficulty starts whenever the average person discovers their belief system disagrees with a scientific theory such as Einstein's Theory of Relativity, The Big Bang Theory or Darwin's Theory of Evolution. When this sort of thing happens, their usual response is to say something along the following lines: 'Well, it's only a theory.' What they mean is that because these ideas are *just* 'theories', they haven't yet been 'proven' beyond all reasonable doubt. There's still some wriggle room. As a result, there's enough doubt about these scientific ideas that there's a valid case for holding a belief that contradicts them.

Nothing could be further from the truth.

The problem stems from the word 'theory' and the fact that it has two quite distinct meanings. What makes this situation worse is that these two meanings are almost exactly opposite. This state of affairs is nothing new for the English language as it is riddled with words we call homonyms — words that are spelt and sound the same but which have different meanings. An example would be the word bow. We use a bow to play a violin, which is very different from the bow we tie in a ribbon around flowers, and that's once again quite different from the type of bow and arrow William Tell used to shoot an apple off someone's head.

It's usually quite clear which type of meaning we're using with homonyms. No one would get confused and think William Tell used a violin bow to somehow shoot an arrow, let alone use a nicely tied ribbon. But the two meanings of the word 'theory' cause us much confusion and grief because this distinction is not clear.

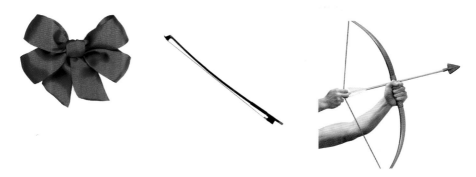

In everyday language the word theory is really little more than an idea or a guess. For example, suppose you walk into your kitchen and discover the pack of sausages you had thawing on your bench is missing and a glass of milk has been knocked over and spilt all over the kitchen floor. You might immediately develop a theory that your cat has jumped up on the bench and made off with the sausages. At this stage it's nothing more than an educated guess, and a pretty good one at that, but you can't yet prove it. On closer inspection you see a trail of paw prints through the spilt milk leaving the scene of the crime. Good evidence supporting your theory but still you can't be sure it was the cat. After all, the cat might have just innocently happened to walk through the milk after someone else had spilt it.

So you follow the paw prints into your study and, sure enough, there is your cat eating the remainder of your sausages. This is now very good evidence for your theory. It's becoming more than just a guess or a wild idea — you've caught your cat 'red handed'. But just to be sure, you go back into your kitchen and replay your home security video surveillance camera. In the first few frames you see the sausages and milk sitting happily on the kitchen bench and then the cat jumps up, grabs the sausages and, in his haste to get away, knocks over the milk with the

trail of sausages. You're now very certain of the culprit. With all this evidence you wouldn't say to yourself, 'Well, I'd better not smack the cat for stealing the sausages because, after all, it's only a theory he stole them.'

At some point you decide the evidence is so overwhelming you're entitled to say, 'For All Practical Purposes (FAPP), I'm sure the cat stole the sausages.' It's no longer a theory.

THE ARGUMENT FOR THEOREM

Here's where things get a bit messy and we have to blame the history of Science for this state of affairs. When scientists talk about a theory such as Einstein's Theory of Relativity, they're not talking about an unproven guess or a hypothesis. Quite the opposite. They're talking about an idea they're at least 99.999999 per cent sure about. Of course nothing in Physics can ever be proven to 100 per cent certainty just as we can't be sure gravity won't stop in a minute and we'll all fly off into space. But it is so unlikely we don't usually worry about it.

Unfortunately, scientists have historically made the big mistake of routinely using the word theory to cover two entirely different levels of certainty:

1. An untested idea or hypothesis which sounds good but which has yet to be proven.
2. An idea which we now have *so* much evidence for that we *must* accept it as FAPP true.

No wonder the average person is confused by scientists' use of the word theory.

It turns out we can all be far more confident of Einstein's Theory of Relativity than we could ever be about your cat because we have so much more independent evidence from thousands of completely unrelated and verifiable sources supporting Einstein. Not only that, but his theory interlocks beautifully with thousands of other mathematical descriptions of the Universe we rely on every day. Mathematics underpins our microwave ovens, GPS navigation systems and computers. Even the slightest, most minuscule error in the smallest detail of Einstein's theory and these machines and equipment wouldn't work.

Fortunately, mathematicians have a much better system for describing ideas than simply the single word 'theory'. When someone first proposes an idea that sounds good but can't yet be proven, they call it a hypothesis, theory or a conjecture. However, once it has been proven to be correct (and of course you can do that with 100 per cent certainty in Mathematics) the hypothesis is then elevated in status and called a Theorem. Notice the different spelling.

The trouble for Einstein was that his idea quite rightly started off as a 'hypothesis-type theory' in 1905 — a brilliant idea with only a small amount of supporting evidence — but after all the overwhelming evidence flowed in it was never elevated in status to a Theorem we can all trust and believe in. The reason for this confusing state of affairs is because physicists are a pedantic lot and don't think the word Theorem should be applied to anything other than Pure Mathematics where you can prove something with 100 per cent certainty. For mathematicians, even 99.999999999999999 per cent certainty isn't enough. But that level of pedantry leads everyone else in the world to think it's still an unproven 'guess' or 'idea' when nothing could be further from the truth. You can be even more sure of Einstein's theory than you can be of your own name.

Because this confusion stems from a problem with the meaning of words I like to suggest a solution. I'd like to propose that once there is overwhelming evidence for a scientific theory it should officially be called a Theorem or a law. We should really say, Einstein's Theorem of Relativity. I'd like to challenge scientists to start using this new nomenclature because it would stop a lot of confusion with theories that are just 'ideas'. After all, the whole reason for language is to help us to accurately convey ideas from one person to another and if using one word (theory) for two meanings is confusing, then we should change the convention. Mathematicians will just have to get used to scientists using their special word Theorem as well. It's going to cause much less confusion this way around than the way it currently is, even if a physicist's definition of Theorem differs from a mathematician's by 0.000001 per cent.

Ideally, there should be an august committee like the Nobel Committee that officially certifies when a major scientific idea changes status from theory to theorem. It's a bit like what happens when a manufacturer makes a very accurate watch. They can apply to the Swiss authorities and get it officially certified as a chronometer. Then it's no longer just a watch, it's a chronometer! And to make matters transparent, there should be a nice record documenting all the evidence supporting this change of status from theory to theorem that anyone can access — written in plain clear everyday language anyone can understand.

This might sound like a bit of fanciful thinking but Science is far too important to be left to the scientists to document and then disseminate to the general public. Think about it for a moment — if we left it to the artists to safeguard works of art most would have been lost long ago. Masterpieces are in much better hands when they're looked after by professionals and displayed in art galleries and museums around the world. Maybe the same applies when it comes to the dissemination of scientific knowledge.

WHAT ABOUT TRUTH?

By now it's obvious we're starting to bump our heads into the whole question of 'Truth'. Is scientific knowledge any different from religious knowledge? Can it be trusted? Is there even any such thing as Absolute Truth? To address these questions we'll need to begin construction of our Compass of Truth in Chapter 7, but before we do, I'd like to tell you about my most memorable experience as a child and why the Compass of Truth is such a powerful and magical instrument.

CHAPTER 6
THE SECRET CODE

'The truth is far more marvellous than any artists of the past imagined it. Why do the poets of the present not speak of it?' — Richard Feynman

LARRY KING LIVE

Imagine one day finding an ancient leather-bound book hidden in a cave. You open it up and find it's filled with secret codes. Not understanding the symbols and wondering how valuable it might be, you take it to a university to see if anyone there can decode it. It takes a full eight months before a small group of cryptologists finally manage to decipher the first page. When they do, they're immediately shocked because each paragraph reveals in exquisite detail things ancient humankind could never have dreamed about. And yet there they are, written down in a book that was proven to be thousands of years old. Its worn-out pages not only foretold the future but also unlocked the deepest secrets of life and the Universe.

Further study revealed the code was even more beautiful than it first appeared. If you read the book in reverse by starting with the last word of the last page and read towards the front of the book with all the words in reverse order, you found each sentence still made perfect sense and revealed additional new truths not spelt out when the book was read normally. Indeed, the book was so complex that some devoted their lives to studying it and became known as Masters of the book. Some even became Grand Masters. The wisest of all Grand Masters were given the most exalted title of Wizards, for they possessed great power and knowledge from their study.

Yes, the idea of a secret code that unlocks extraordinary knowledge is why we love books like *The Lord of the Rings* or *Harry Potter*. If such a book really existed it wouldn't be long before every TV network in the world was begging you to appear on their show.

But what if there really were books like that and there really were great Wizards alive today?

A YOUNG BOY

When I was about seven my dad gave me a series of *Science Encyclopaedias for Young Children* which I read with the interest of a wide-eyed boy discovering new things about the world. Actually, I didn't just discover new things, but whole new categories of things I never knew existed. There was a volume on Biology which introduced me to all sorts of weird and wonderful animals: dinosaurs of impossible proportions and ferocity, now thankfully long gone, and animals whose physiology was so different they didn't even have lungs or stomachs like ours.

While that book was interesting, it didn't really grab me because it seemed to be little more than a glorified catalogue listing exotic items found in nature. Likewise, there was another book on Chemistry which told me how to combine one chemical with another to produce explosions or vivid clouds of gas. All very amusing, but it still seemed like a collection of facts and observations.

But there was one more book in the series, and as soon as I picked it up I knew it was special. This book didn't just catalogue observations and label objects — it made statements it claimed were Absolutely True. Statements which apparently not even God in all his majesty and power could alter, because they were that pure. Statements uncontaminated by any human limitations and completely independent of our human frailties. This total and absolute purity struck my young mind as being incredibly special. I liked things being organized and clean and, according to this book, the statements inside were the cleanest and purest things in the Universe.

Imagine my delight, then, when I discovered these statements weren't just Absolutely True, they also unlocked the secrets of the Universe at the same time. Yes, this book was about Mathematics and Physics. Its pages introduced me to many secret formulae — each of absolute clarity and infinite precision, each able to be harnessed by a Wizard of sufficient skill to discover totally new and unexpected things about how the Universe works.

This ability to master Mathematics has produced many of the greatest advances in human history. It helped drag us from the cave where the average lifespan was only 25 years spent in a grinding struggle to get enough food to survive, to our modern comfortable homes packed with electronic gadgets. In case this sounds like a wild claim, let's consider a young patent clerk who sat quietly behind his desk 100 years ago manipulating Maths symbols without any experimental equipment or any laboratory to help him — simply decoding the 'Great Code of the Cosmos'. Codes which allowed him to single-handedly uncover the mechanisms that would later allow us to produce MRI scanners, laser beams, computers, nuclear weapons, DVD players and GPS — to name but a few.

Yes, Einstein was one of the true Wizards in a long line of Wizards like Newton or Plato before him. In fact he was so advanced it would take the rest of us another 100 years to fully realize what he discovered when he turned his equations first this way then that. Each new twist unleashed the most unexpected and mind-boggling discovery — just like our book of codes which can be read backwards or forwards.

It's not surprising that reading this book left more of an impression on me than if I'd actually discovered that ancient book of codes hidden in a cave. It motivated me to win the Senior Prize in Mathematics at university many years later and has given me more moments of wonder and awe than I could have ever imagined.

SPELLING VERSUS READING

At this stage you may be thinking to yourself, 'Well, Maths was never like that for me. It was just a pile of boring old formulae.' And you're probably right. The best way to explain this is to remember what it was like when you were learning to spell. C-A-T spells cat. D-O-G spells dog. And so on. It was all pretty boring stuff, wasn't it? But once you've mastered the alphabet and learnt how to spell and read — why you can suddenly read novels that transport you to the other side of the world or inside someone else's mind.

Unfortunately, most school Maths is like spelling or learning the alphabet and it's taught by teachers who themselves can only 'spell' and have never learnt to 'read' Maths. This is a great tragedy. And yes, Maths *is* hard — just as you would expect it to be hard to decode that dusty book of ciphers hidden in the cave — but surely, most worthwhile things in life *do* take a bit of effort. And if you personally don't have the resources to climb that mountain then the good news is that others have done the hard work for you and are happy to report what they see from the summit. In this way you needn't let your own lack of ability stop you from seeing the view. Of course it won't be quite as exciting and fresh as if you saw it first-hand and many scientists are not very good at describing the view in a way that is even remotely interesting. I hope this book helps a little in this regard.

The other point to realize about Maths is where it's useful and where it's of no value at all. Maths and Logic are perfect tools for sorting out questions of Truth and Falsity but they're hopeless at addressing issues which are neither true nor false; things such as happiness, music, love and the arts. These things add meaning and richness to our human existence and it's here that 'words' come into their own. So you need both Maths and words, each with their own unique contribution.

COSMIC RELIGION

Learning how to 'read' the Universe in Mathematics (instead of just 'spell' with it) is without doubt one of the most profound experiences anyone can ever have. For Einstein, this process was so moving he described it as the purest 'religious' feeling possible.

By now you may be wondering, 'Why Maths? What's so special about formulae? Surely some people think in formulae and others prefer words and neither should have preference over the other.' The answer is that natural languages such as French or English are just too clumsy to describe the essence of the Universe because they lack the correct structure and discipline. They completely miss the very thing you are trying to capture. The best way I can describe this linguistic deficiency of normal languages is to say it's like trying to describe the experience of sex to a virgin using only the language of motion and position. You just can't do the experience justice with such a limited and awkward vocabulary. But with Maths it appears we're using the native language of the Universe itself, allowing its raw unadulterated beauty and symmetry to be laid bare. We might even say mathematical 'words' are the very fabric from which the Universe itself is woven.

It is that direct and that pure.

> *'I marvel at the unreasonable effectiveness of mathematics in describing our Universe.'* — Eugene Wigner

Another mistake is to think Maths is just about numbers. This is another hangover from your school days because advanced Mathematics hardly ever uses numbers but instead makes extensive use of symbols. Just think of the most famous equation in the world, Einstein's $E = mc^2$. It has only three symbols and nothing else: E (for Energy), m (for matter) and c (for the speed of light).

So why is Maths special and what's the point of all these mysterious symbols?

Maths symbols are simply shorthand ways of writing down complex sentences or ideas so we can manipulate them more easily. At some point we realize we're going to use an idea or sentence over and over again, and so rather than writing that idea out repeatedly, we give that concept a special label such as λ. Doing this means we can combine and relate ideas without losing track of the logic. The reason we need to do this is because our brains are just too small. We can't hold more than seven complex ideas in our brain at one time, so when we get to complicated things we need to simplify them with symbols and write them down so we can see how one idea follows from another.

What's really exciting about doing this is that you start with relatively trivial things you already know and simply by 'following the logic' you end up in places you had no idea existed. You discover things beyond your wildest imagination which turn out to be *true*! Who would ever have thought you could turn a microscopic piece of metal (uranium) into a huge fireball (an atomic bomb)? That was just one of the unexpected results of $E = mc^2$.

> *'How can it be that mathematics, being after all a product of human thought which is independent of experience, is so admirably appropriate to the objects of reality?'* — Albert Einstein

This brings me to the essence of what I like so much about Maths and formal Logic. It's not the Maths itself that is exciting, but rather where it takes you and what it allows you to discover. The best way I can describe this is to think of a telescope. I could never get excited by a physical telescope or the lenses and mechanics that make it work, but when I look through one I'm able to see planets such as Saturn with her beautiful rings or millions of galaxies, each containing billions of stars spread across unimaginable stretches of space. It's what the telescope does that makes it precious. The same goes for Maths.

'It is very difficult to elucidate this "cosmic religious" feeling to anyone who is entirely without it. ... The religious geniuses of all ages have been distinguished by this kind of religious feeling, which knows no dogma and no god conceived in man's image; so there can be no church whose central teachings are based on it. ... In my view it is the most important function of art and science to awaken this feeling and keep it alive in those who are receptive to it.' — Albert Einstein

CHAPTER 7
THE COMPASS OF TRUTH

'The trouble with the world is not that people know too little, but that they know so many things that ain't so.' — Mark Twain

THE GOLDEN COMPASS

Given our human frailties, and we only scratched the surface in Chapters 1 and 3, it would be brilliant if we could find some sort of instrument that allowed us to tell whether any idea or theory was *true* or *false*. Anyone who possessed such a device would obviously wield amazing power because they'd have knowledge other mortals could only dream of. This idea was central to the first book in Philip Pullman's fantasy trilogy *His Dark Materials*, which was later made into a film called *The Golden Compass* starring Daniel Craig and Nicole Kidman.

In that movie, the Golden Compass resembled a normal compass but with mystical symbols inscribed around the outside of the face. To operate it you simply asked the Golden Compass a question and the needle swung around and pointed to one of the symbols indicating the level of Truth behind an idea.

In this chapter we're going to develop our own Compass of Truth. It will become a valuable instrument you'll be able to use to check your own beliefs so you don't end up 'playing life by the wrong rules'. You'll also find it comes in handy when you are having debates with friends on contentious topics.

Rather than just showing you the finished product we're going to build it up step by step so you know what's inside it and how it was constructed. That will give you confidence in its workings and increased mastery when you use it. It's a bit like pilots having greater confidence in their plane when they know every nut and bolt that holds it together.

THE THREE TYPES OF STATEMENT

Our starting point is to realize we can classify all statements into one of three completely distinct categories:

1. Statements that are *true*.
2. Statements that are *false*.
3. Statements that are *neither true nor false*.

We can place these three categories on a diagram and when we do we notice they don't overlap. Because they are completely distinct I've made them different colours.

It may sound like a wildly arrogant claim to say *every* statement must fall into one of these three categories, but it's true. A good analogy is to think about what happens when we compare one person's height to another's: we also find there are exactly three distinct possibilities. For example, suppose I compare Susan's and Mary's heights at precisely 9 a.m. this morning, then whatever the heights of those two girls are, we must also get exactly one of the following three answers:

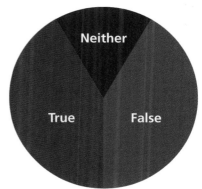

The Truth Landscape

1. Mary is taller than Susan.
2. Mary is shorter than Susan.
3. Mary is neither taller nor shorter than Susan (she is the same height).

It's vitally important to stress at this point that we're *not* saying we can *ever* tell which of those statements is true. We may never be able to measure their heights accurately and so we may never know who is taller or whether they're the same height. All we're saying is that *one* of those statements must reflect Reality and *two* of them must not. There are no other possibilities and we can't have two of those statements ever being true at the same time. For example, Mary can't be both taller *and* shorter than Susan at the same time. We're also not saying they should always be true forever because one of them may grow faster than the other. Nevertheless, at any instant in time there's only one possibility, even if we can never work out which one it is because of our ant-like limitations.

THE NEITHER CATEGORY

Now we all know what we mean when we say a statement is true or false but what on Earth do we mean when we say a statement is *neither true nor false*? Let me give you a trivial example of a 'Neither' statement:

'His height was 75 kg.'

This sounds like a sentence that *could* be true because it has the same shape and form as the following perfectly valid sentence:

'His weight was 75 kg.'

But just because a sentence has the right shape and form this doesn't mean it might ever be true or false. In fact the first sentence is nothing more than gobbledegook because you can't measure height in kilograms. So, despite the fact that it looks like a sentence, 'His height was 75 kg' is nothing more than a meaningless jumble of words.

There is also another type of Neither statement which the following sentence illustrates:

'Picasso was a better artist than Van Gogh.'

The problem with this sentence is that unless we can decide unambiguously what the words 'is a better artist' really means, then we have no way *in principle* of ever deciding whether that statement is true or false. Our difficulty stems from the fact that there's no clear-cut definition of what the phrase 'a better artist' means — at least not as far as the objective Universe is concerned. Rather, it's a subjective statement. I might think Picasso was better while you might Van Gogh was, and we'd both be entitled to our opinion. I could modify this statement into one which would be either true or false by saying something like:

'I *personally* think Picasso was a better artist than Van Gogh.'

Because I've now described the phrase 'a better artist' in terms of my own personal preferences there's a possibility this new statement is either true or false. We could also get around the problem by defining the phrase 'a better artist' in terms of some objective measure such as 'who was the most popular'. Then we could decide if our statement was true or false by simply counting up how many people prefer each artist. The only conclusion we can come to is that until we define exactly what we mean by the phrase 'a better artist', our original sentence is neither true nor false.

It turns out there are a huge number of statements that sound like they could be either true or false but which are really Neither statements. We'll need to be on high alert for them because they're slippery customers and cause a great deal of trouble and are often at the centre of many a heated debate without people realizing it. But this doesn't mean these Neither statements lack any value in our communication — quite the opposite, as there are many valuable things in life such as art and music that aren't measured in terms of Truth. Listening to Mozart is a wonderful experience just as falling in love is.

THE GREAT UNKNOWN

It turns out there is one more possibility we haven't yet mentioned, which I call the Great Unknown. This region contains all those things which are beyond our knowledge. To go back to the ant and Ferrari analogy we used at the start of this book, it represents everything hidden beneath our hood.

In the words of Donald Rumsfeld, the Great Unknown will contain two types of things:

1. Things we know we don't know.
2. Things we don't even know we don't know.

An example of the first type of statement (option 1 above) might be something like:

'There is two-legged life on other planets that wears brown shoes.'

This rather silly statement must either be true or it must be false because it makes a genuine claim about what's happening in the Universe. There either is, or there isn't, a two-legged creature with brown shoes on another planet. But the point remains, we just don't know either way because it's beyond our current sphere of knowledge. Nevertheless we do realize we don't know the answer to that question. There's also a vastly larger list of things we don't even realize we're ignorant about. For example, the ancient Romans had no knowledge of electrons or viruses so they wouldn't have been able to even form a sentence with those words in them.

We could represent the Great Unknown by drawing a huge White region outside our existing diagram, but if we did, our diagram wouldn't be to scale because with our ant-like knowledge, the things we don't know will be billions of times larger than the things we know. In all likelihood the White circle would extend off the page for many kilometres.

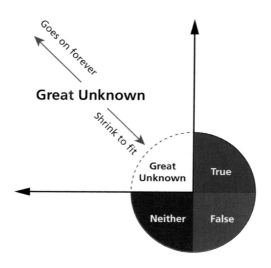

In order to simplify things I'm going to shrink the Great Unknown and bring it inside and put it next to the other three regions.

This now gives us a much more manageable diagram.

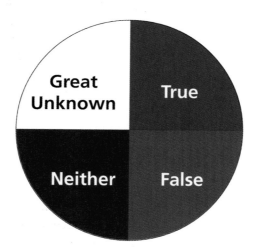

Now it's all very well talking about what's True, False or Neither, but we haven't yet talked about what we *believe*, and that after all is a key part of what this book is about.

We might:

- believe in true things — in which case we'd be *right*
- believe in things that are false — in which case we'd be *wrong*.

On top of this, sometimes we'll be very sure of our beliefs while at other times we might be quite uncertain of them. We can represent these varying levels of certainty by having different shades for our colours on our diagram. When we're *very* sure we'll have a solid rich colour and when we're not sure the colours will be faded out. This gives us the next diagram which is starting to look more like our completed Truth Compass.

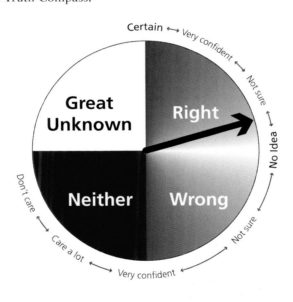

THE POINTER OF CERTAINTY

You'll see I've also put in a pointer which indicates what our level of belief is. If the pointer is at the top (12 o'clock) this means we're both totally confident of our belief and we're *right*. For example, a belief placed near the top in the Green region might be something like, 'If you cut off a person's head they'll die.'

As the pointer moves clockwise we become less and less sure of our beliefs until at 3 o'clock we admit we simply have no idea whether we're right or wrong. You'll notice the green also becomes lighter and lighter as we go further around the clock to match our diminishing confidence. At 3 o'clock the green is completely faded because we now have absolutely zero confidence in our belief at that point. What's important to remember is that everything in the Green region must actually be *true* regardless of how certain we are of it. Of course, we can't yet tell what is actually true or false — that's going to come later — all we are doing here is showing where our beliefs would be if we did know whether they were true or false.

As soon as we go below three o'clock we venture into the region where we believe in something but now we're *wrong*. Beliefs placed just below three o'clock are all those false ideas we believe in but with very low confidence. This makes sense because our level of confidence just above and just below three o'clock hasn't changed much; the only difference is that now we're wrong. As we continue to move around the clock we get more and more confident in ourselves but unfortunately despite our increasing confidence it turns out we're still actually wrong. Being wrong and confident is the most dangerous territory which is why I've coloured it in red. It's where we'd put ideas like: 'Sleeping with virgins cures AIDS.'

THE SANTA CLAUS EXAMPLE

To show how the right side of the Truth Compass works let's consider my childhood belief in Santa Claus.

When I was a young boy my parents told me about Santa and I always trusted everything they said with 100 per cent certainty. Sure enough, when I went down to the fireplace each Christmas morning I found a stack of nicely wrapped presents and a card from Santa with my name on it. With solid evidence like this I was totally and utterly convinced '1. Santa is real.' We know Santa doesn't exist which means my childhood belief was both confident and wrong and so it starts out very close to six o'clock.

As I grew older I started to have my suspicions. How could Santa get around the entire world in one night? Why did Santa's writing on my Christmas cards look the same as my mother's? Slowly but surely my confidence in Santa began to slip and so my pointer of confidence moved backwards from six o'clock to five o'clock to four o'clock, and finally when I was totally unsure, to three o'clock.

As more time goes by I notice the presents Santa delivers match my mother's shopping and so I start to have my suspicions that '4. Santa is *not* real.' In other words I begin to have a correct belief, although with very little confidence. Before long my pointer is at two o'clock and then when I finally confront my mother and she admits the Truth, my needle of confidence quickly swings backwards towards 12 o'clock. I'm now very certain: '6. Santa is not real.'

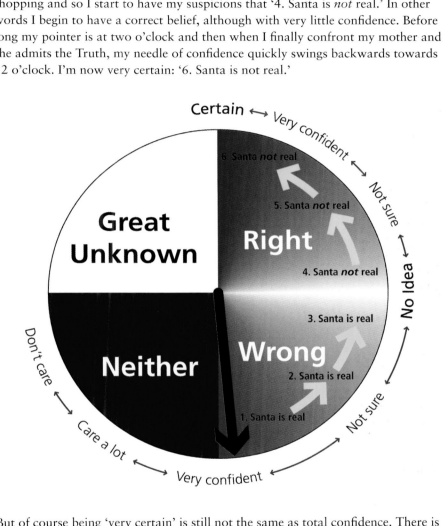

But of course being 'very certain' is still not the same as total confidence. There is still no way I can *prove* to everyone's satisfaction that a real Santa doesn't exist somewhere.

This now brings us to one of the most interesting questions we can ask ourselves: 'Is there such a thing as **Absolute Truth**?'

CHAPTER 8
IS THERE ABSOLUTE TRUTH?

'The highest form of pure thought is mathematics.' — Plato

Can we ever know something with such confidence we can be *totally certain* it is true despite our ant-like limitations? We're right to be sceptical about whether **Absolute Truth** exists because when we look back over history we see that most of what we used to believe in was in fact wrong. From blood-letting for curing fevers to the Earth being flat we're obviously quite fallible. Because our history doesn't give us much confidence it's not surprising to hear the common claim that 'all Truth is relative'. Sometimes this is rephrased as 'human Truth is a mere approximation of Reality'. But is this true?

All we've shown by our past mistakes is that some beliefs were misplaced. No one ever claimed they could *prove* with 100 per cent certainty the Earth was flat or that black swans didn't exist. These were just beliefs we convinced ourselves were true based on our experiences and observations. But being very sure of something is entirely different to *absolute* certainty. Absolute proof is never subject to change or error, which is why it's the Holy Grail.

Where we went wrong with our flat Earth, black swans and Santas was that we relied on observation and evidence. But observation and evidence are never enough to ensure certainty because what we see is always tainted by our human limitations, as we saw in Chapters 1 and 2. So does this mean all is lost? Is there no such thing as absolute knowable Truth after all?

Before we answer that question let's see where we'd put Absolute Truth on our Compass if it did exist.

The obvious place to put it is at 12 o'clock because that's the only place we are both correct and have 100 per cent certainty at the same time. But I'll need to make it a bit wider than just a narrow line if I'm going to have enough room to put things in it. Therefore, I'll take a bit of poetic licence and make it into a small sector so you can see it. I'll also colour it gold because this must be our most sought after 'Holy of Holies', for it contains Absolute Truth we humans can *know*. But if there is something we know is absolutely true, then it's opposite *must* also be absolutely false. This means we need two regions at the top of our diagram: **Absolute Truth** and **Absolutely False** as shown below.

So, is there anything in our Gold or Black regions?

PLATO'S ANSWER

It turns out there are many statements which are perfectly true and which we can prove with absolute certainty. All of Mathematics contains statements which are 100 per cent provably true. They don't require any observation, measurement or interaction with the Universe and this helps free them from our human limitations. Mathematical statements are necessarily true and there is no shadow of uncertainty or doubt about them.

1 + 1 = 2 is always true. It always has been and it always will be.

The truth of this equation follows from the *definition* of the numbers '1' and '2'. It's also why we never have to prove the laws of Mathematics every time we put money in the bank. Every person in the entire world can rely on the fact that if they deposit $50 on Monday and another $70 on Tuesday, the bank should credit them with having deposited $120. We'd be pretty upset if the bank tried to tell us the laws of Maths had changed over the last few days and $50 + $70 now only equalled $35. We'd be outraged if the bank said they now had their own version of arithmetic that was different from ours or that 'we are all free to believe in any version of Maths we like because they're all equal'. It's a silly argument and yet the very same people who argue 'all Truth is relative' are the same people who would be justifiably upset if their bank balances were suddenly relative. You just wouldn't accept it in real life.

Unfortunately, arithmetic like this doesn't add anything new to our view of the world. Arithmetic is true only because we've defined what each of the numbers meant in advance. We've defined '2' in such a way that its meaning is nothing more nor less than being what happens when you combine two '1's. But starting with simple definitions like this and then examining them in more detail leads to some surprising and quite unexpected results — all without relying on any observation at all.

Here's an example:

1. If I *define* a 'right angle' as being ¼ of a circle (360 degrees) = 90 degrees
   then
2. For any flat **triangle** containing a right angle, the **lengths** of its three sides are related by Pythagoras' theorem: $c^2 = a^2 + b^2$

It doesn't matter whether or not I can ever draw a perfect right-angled triangle in the real world; the statement is still *always* true. I don't need to obtain evidence by measuring hundreds of real triangles. I can *prove* it from the meaning of the words 'triangle', 'length', 'right angle', 'flat' and nothing else. What's incredibly surprising to most people is that this relationship ($c^2 = a^2 + b^2$) is hidden inside the two sentences like a secret code waiting to be unpacked. This is part of what makes Maths as exciting as deciphering a book of secret codes like that in Chapter 6. We discover things of Absolute Truth we didn't expect and these discoveries turn out to be amazingly useful — all hidden there in our everyday words.

'One reason why mathematics enjoys special esteem, above all other sciences, is that its propositions are absolutely certain and indisputable while those of other sciences are to some extent debatable and in constant danger of being overthrown by newly discovered facts.' — Albert Einstein

In this way Mathematics and formal Logic are things we can know with absolute certainty to be true. This means we can put all mathematical statements in a special Gold region at the top of our Compass of Truth. Mathematics is therefore rightly called the Queen of the Sciences because it is the *only* science that allows us to know things with absolute certainty. Physics, Chemistry and Biology can never claim such an exalted status because they rely on observation and experiments. This is why the great Greek philosopher Plato had the following words inscribed over the entrance to the teaching academy he established in Athens:

'Let no one ignorant of Geometry [mathematics] *enter here'* — Plato

TAUTOLOGIES

There are other statements we can place inside the central Gold region of Absolute Truth as well. The statement 'An adult human is older than a baby human' is such a statement if we accept the normal meaning of the words. The reason this statement is always true is because it follows directly from the definition of the words 'adult', 'older' and 'baby'.

These sorts of statements are called tautologies because they don't tell us anything new — even if at first glance we might think they do. All they do is re-establish the definition of what we mean by the words we're using.

LOGIC

Logic is like Mathematics and is also always true. Here is a combination of three statements that are true regardless of what they refer to:

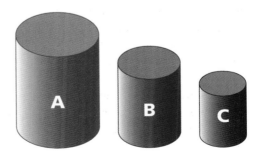

1. **If** A is bigger than B.
2. **If** B is bigger than C.
3. **Then** A is bigger than C.

Again, the interesting thing about these three statements is that we're not saying there's anything in real life that corresponds to these items A, B and C. All we're saying is that *if* these first two statements are true (and of course they might not be) *then* the third statement *must* be true.

Namely, A is bigger than C.

Logic like this allows us to establish connections with things that weren't immediately obvious. Just as it wasn't immediately obvious that $c^2 = a^2 + b^2$ for triangles so too formal Logic allows us to discover new information based on existing information.

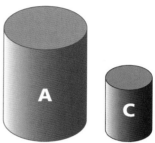

> *'Pure mathematics is, in its way, the poetry of logical ideas.'*
> — Albert Einstein

SELF-EVIDENT TRUTH

This brings us to a rather startling point in our discussion. The rules of Mathematics and Logic are called **self-evident** because they are **true out of necessity**. Not even God can break or bend the rules of Mathematics or Logic and no one needs to invent them. They exist there all by themselves. All we can do is discover them in the same way an explorer discovers a new mountain or island. In our example of the red, green and blue cylinders above, it is simply impossible for the third statement to be false if the first two statements are true. The combination *must* be true because not being true is impossible. In the same way all of Mathematics *must* be true. Pythagoras' relationship $c^2 = a^2 + b^2$ would be true even if no one had ever drawn a triangle or the Universe itself never existed! We might not realize it was true — we might never discover it — but it would still be true because it is impossible for it not to be.

This idea of something being *'true out of necessity'* is a topic we'll come back to later when we consider what caused the Big Bang in the first place. But just imagine for a moment what it would mean *if* we discovered the Big Bang was as 'self-necessary' as Maths or Logic. If it *had* to happen all by itself — without any intervention or agent causing it in the same way no one needed to invent the logic of the cylinders or Pythagoras. If the Big Bang was true out of necessity. If the Universe *had* to exist. That's an idea that's probably both scary and exciting at the same time.

'One cannot escape the feeling that these mathematical formulas have an independent existence and an intelligence of their own, that they are wiser than we are, wiser even than their discoverers.' — Heinrich Hertz

ABSOLUTELY FALSE AND ABSOLUTE TRUTH REGIONS

Just as we can prove with 100 per cent certainty that $c^2 = a^2 + b^2$ is true, so too we can also prove with 100 per cent certainty that $c^2 = 2a^2 + b^2$ must be absolutely false. This is why we needed to add the Black sector of Absolutely False to our Compass of Truth. We need to place it right beside Absolute Truth because everything inside it are things we can prove to be false with absolute certainty.

THE REAL WORLD AND FAPP

By now you've probably spotted a pattern with Mathematics, Logic and tautologies — they refer to abstract things that may or may not exist in physical Reality. This raises a problem. They may be true but they may not describe anything in the real physical world. For example, Mathematics only concerns itself with abstract things such as 'perfect circles'. Unfortunately, there are no such things as 'perfect circles' in the real world. We only find circles that are almost perfect. Einstein summed up this rather frustrating situation as follows:

'As far as the laws of mathematics refer to reality, they are not certain; as far as they are certain, they do not refer to reality.' — Albert Einstein

There's an almost match between Reality and Mathematics — but we can't quite make the match absolutely perfect.

But this situation shouldn't worry us too much. Even though we can't draw a perfect right-angled triangle this doesn't mean we should give up and throw in the towel. If I take a protractor and draw my best attempt at a right-angled triangle, I can be pretty confident the three sides are going to be very close to $c^2 = a^2 + b^2$. And the more accurately I draw my triangle the closer my answer will be. While my measurement in the real world is not *absolutely* correct, it is so close I can say that For All Practical Purposes (FAPP) right-angled triangles have sides whose lengths are given by $c^2 = a^2 + b^2$.

This idea of FAPP is an incredibly powerful concept. It covers statements that are so overwhelmingly close to describing the Universe that even though we can't prove them to be absolutely true we'd be completely stupid if we didn't accept them as being true For All Practical Purposes because they are so closely related to things inside our Gold region. Perfect triangles (which don't exist in Reality) always follow Pythagoras' formula and so are placed in the Gold region of Absolute Truth while 'real triangles' are so close we don't need to keep measuring their lengths all the time.

We can rightly place 'real triangles' right on the boundary of the Gold region. We can have an extremely high confidence in real triangles and their lengths precisely because they're directly related to things inside our Gold region. So we need to place FAPP right next to Absolute Truth, which is where you'll see I've already sneakily placed it.

Not surprisingly, there's a huge body of 'knowledge' we rely on each day that's also very close to the Gold region. For example, I don't wake up each morning and clutch onto my bed in case gravity has suddenly failed. It's reasonably certain that gravity is caused by the mass of the Earth, and if I wake up in bed in a normal fashion, it's surely reasonable to assume the Earth is still somewhere around and gravity is going to do its thing. Of course I can't prove gravity will always be here. One day the Earth will disappear when the sun dies and swallows it, but all things being equal I don't have to worry about this each morning I wake up.

Similarly, I can be fairly certain a person will die if we cut off their head. It's always been that way and probably always will be, but notice I'm starting to get a little less certain as I go on.

I was:

- totally confident about abstract perfect triangles (which don't exist)
- very confident about real triangles (which do exist)
- a little less confident about gravity
- and now I'm slightly less confident about people without heads dying.

The reason for this progression is because the further away I get from the Gold region and the less connections I have to things in the Gold region the less confidence I have. Real triangles are very closely related to perfect triangles in the Gold region and gravity is supported not only by my daily observations but also by a vast interlocking network of Mathematics that's also in the Gold region. But gravity is a little further away from the Gold region because it relies on more assumptions than real triangles do.

And as far as people without heads go, I'm even less certain because cockroaches can live for months without their heads and so it might just be possible that humans could do so as well. But it's extremely unlikely because I know how humans breathe and how our lungs provide oxygen to our brains. Cockroaches on the other hand don't have lungs and their legs aren't really controlled by a brain like ours. I can see there is a difference between cockroaches and humans and for that reason I'm not concerned I'll see people without heads walking around my supermarket tomorrow.

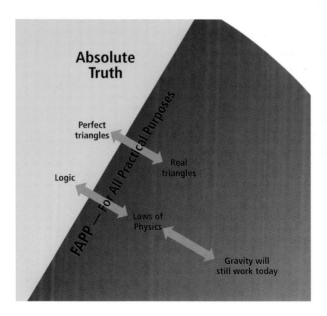

'The laws of nature are but the mathematical thoughts of God.' — Euclid

WHICH THINGS CAN WE HAVE CONFIDENCE IN?

This brings us to a key point. The reason why I can have high confidence in some observations is because they're closely connected to things in the Gold region of Absolute Truth. Other beliefs or observations have a more tenuous connection with the Gold region and so I should be more wary of them. That's why I was so easily fooled into believing in Santa Claus when I was a child, because Santa Claus has *no* connection with anything in the Gold region of Absolute Truth. But the belief the Universe began with a 'Big Bang' does have thousands of connections with items in the Gold region and with other items I trust on the boundary I call FAPP.

The truly amazing thing is how well connected mathematical objects in our Gold region are with real objects in our daily physical lives. This is why Science has been the prime force that helped humans to drag ourselves out of our wild animal state. In only a few thousand years we went from spending our entire lives scratching in the ground for food and dying of diseases before we turned 20, to flying planes around the world and living lives of relative luxury. We have our understanding of Absolute Truth and FAPP to thank for the comfortable and healthy lifestyles that billions of people in the 'first world' enjoy today — something we call Science.

'If there is a God, he's a great mathematician.' — Paul Dirac

CHAPTER 9
GENESIS REVISITED

'Any religion which cannot survive a collision with the truth is not worth worrying about.' — Arthur C. Clarke

THE BIBLE'S CLAIMS

As we noted back in Chapter 4, over half of all Americans believe God created the Universe as described in Genesis. The question then is: can the Bible's account of creation be reconciled with the overwhelming evidence for the Big Bang? In other words, where should we place Genesis on our Compass of Truth?

The answer to this question has consequences that go far beyond deciding how our Universe began and reaches out to touch every corner of our daily life. It can literally determine the difference between who lives and who dies, as we'll soon see, which is why we need to take this question very seriously indeed.

We start by looking at our Compass of Truth to see which region the Genesis account should lie in. Because we know about the story it can't lie in the White region of the Great Unknown. It also can't be placed in the Black region of Absolutely False because the Genesis account makes specific claims about the real world. So far there is no self-contained mathematical proof Genesis is wrong. That only leaves four regions:

1. Absolute Truth — Gold region
2. Neither — Blue region
3. Wrong — Red region
4. Right — Green region.

Next, we need to determine exactly how we should interpret the Bible, otherwise any comparison between the Bible and the Big Bang would be invalid. The three options we have available to us can be broadly classified as:

1. The literal word of God, which should be taken at face value as a true representation of Reality.
2. An allegory inspired by God to illustrate a key point.
3. A man-made myth that has no divine authority behind it.

Let's clarify each option before we decide which is correct.

1. Literal word of God

We start with the claim the Bible is the direct infallible word of God. Even though the Bible was written down and recorded by people, the words in it were nevertheless inspired by God who revealed the exact message he wanted conveyed to his prophets. The prophets merely acted as scribes. There are no 'errors' in the Bible and the words in it accurately reflect God's direct intentions and message to us.

We should therefore interpret the Bible exactly as it was written in a normal everyday manner. We can also reasonably expect God did not deliberately set out to deceive us. In this case, because God 'knows everything' we would be justified in placing Genesis in our Gold region of Absolute Truth even though we can't prove it ourselves.

2. Allegory

While the words in the Bible were still inspired by God, it's written using allegories and lessons that should not be taken literally. Rather, the stories and passages illustrate important points through stories and fables. Despite that, the Bible is still more than a collection of humanity's own invention and self-acquired knowledge. The Bible still occupies a unique position because it bears the hallmarks of divine inspiration and ultimate Truth. In this case Genesis should be placed in the blue Neither category of our Compass of Truth.

3. Man-made and open to error

The Bible is a collection of lessons and stories that were invented purely by people. There is no divine revelation. Nevertheless, these man-made stories may still contain worthy lessons and important concepts that can add value to life and provide us with guidance. But the Bible has no special status compared with other collections of human wisdom. Because such wisdom has been shown over the years to be riddled with errors, it may well be that the Bible also gives false advice from time to time. In this case, if Genesis can be shown to be false, it would lie in the red Wrong category of our Compass of Truth.

Each of these three options fits in a different place on our Compass of Truth as the diagram below shows.

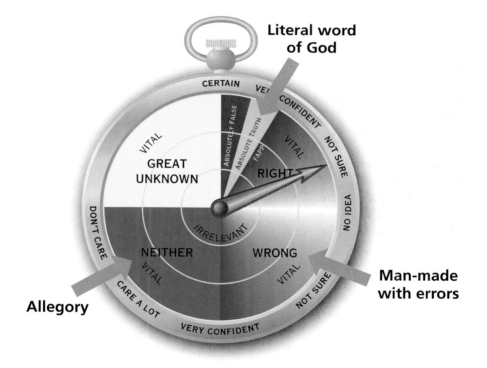

Let's take each option in turn and see if the Bible and the Big Bang can both be true at the same time.

OPTION 1: THE LITERAL INTERPRETATION — A DAY AND A NIGHT

If we use a literal interpretation of the Bible we see that Genesis gives the following timeline for the creation of various objects in the Universe:

Day 1: God created light

Day 2: The water above (clouds and rain) and the water below were separated by Heaven

Day 3: God gathered the water below together so dry land appeared
God called them 'seas' and the dry land 'Earth'
God created grass, herbs and trees

Day 4: God created the stars and put them in the sky
God created the moon and the sun

Day 5: God created all the fish and birds

Day 6: God created all the animals on the land
God created man and woman

Day 7: God rested.

This timeline immediately comes into direct conflict with the Big Bang on a number of fronts. The first is obviously the amount of time between the start of the Universe and the formation of the stars, planet Earth and animals. According to our previous measurements of speed and distance in Chapter 4, it's clear this process took billions of years before the Universe expanded sufficiently so it cooled down to temperatures where life was possible. (Remember back to the analogy of the cup of coffee cooling down.) In the Bible this was supposed to have taken only six days.

The most common reply given by Christians who believe in a literal interpretation of the Bible when confronted by the evidence for the timescales of the Big Bang is to quote 2 Peter 3:8 which says, 'A day with the Lord is like a thousand years, and a thousand years are like a day.' Using this logic they claim that God's time frame is different from our human time frame and so there really is no conflict after all. Or to put it another way, that the word 'day' really only refers to some 'unspecified time period'.

Unfortunately, this line of argument fails for a number of reasons. Firstly, instead of thousands of years we actually need not only millions of years per day but billions of years. But let's put this very minor objection aside and move on to more serious ones.

In Genesis it also says that on day four God made the sun and the moon so that the sun ruled over the day and the moon ruled over the night. Any sensible interpretation of this sentence would clearly mean the 'day' was our normal 24-hour period. How else could we have the sun being visible in the day and the moon being visible in the night if each 'day' was really some unspecified time period? And why would God add that easily understood clarification of what a day was, if this clarification deliberately misled people? To further reinforce this literal interpretation of a day being our normal 24-hour period, the end of each day is finished with the statement:

> 'There was an evening and there was a morning — the third day.'

> 'There was an evening and a morning — the fourth day.'

> ... and so on.

It just doesn't make any reasonable sense to say this chapter of the Bible is compatible with a universe that expanded and cooled over a period of 13.7 billion years. In fact, for almost the entire history of the Christian Church, Genesis was interpreted by all the priests, bishops, popes and devout followers to mean the Universe was created in six literal days each comprising of 24 normal hours. It was only after scientific evidence began to mount that Christians changed their tune and began to claim a 'day' wasn't a day after all.

This means that millions of followers and devout religious leaders were completely misled by their normal interpretation of the Bible for some 2000 years. Hardly a satisfactory result for a book inspired by a God who set out to instruct humanity — particularly when he knew about our human frailties. If any school teacher failed so miserably in getting their message across to their students they would certainly be sacked from their job. If this wasn't bad enough, worse is yet to come.

THE LIFE CYCLE IS WRONG!

Even if it was possible (which it isn't) to use some linguistic wriggling to get out of the problem over the length of the 'day', one thing that absolutely *must* be correct if we're using a literal interpretation of the Bible is to get the order of events correct. Think of it this way. We know monarch butterflies go through four stages: egg, caterpillar, chrysalis and finally the beautiful butterfly. The order *must* be correct or we won't get our butterfly.

We wouldn't hesitate to tell our children they were wrong if they told us it went:

chrysalis ➜ egg ➜ butterfly ➜ caterpillar

or that a butterfly gave birth to a chrysalis or a caterpillar laid eggs. That is just plain wrong!

But this reversal of order is exactly what happens with the Genesis account.

In Genesis, the Earth and the oceans were created on day three while the stars were created later on day four. But this is exactly the opposite of what actually happened in the Big Bang. When our Universe exploded, the stars came first and the Earth followed many billions of years later — for one very good reason. The Earth, and all the elements which make up the Earth, were formed from the leftover remnants that are flung deep into space when a supermassive star dies. These early first-generation stars laid the 'eggs' that gave birth to all the elements such as carbon and iron which go to make up our life. Until these first-generation stars die, these elements are completely missing in the Universe.

Without going into all the details, here's a quick snapshot of what happens. First of all the Universe explodes in a great fireball of the Big Bang which cooks up hydrogen and helium in exactly the manner predicted by our Cosmic Oven (the Large Hadron Collider I mentioned in Chapter 4). After millions of years, the attractive force of gravity slowly clumps together this hydrogen and helium into supermassive stars which are hundreds of times bigger than our sun. After a few more billion years, these supermassive stars will have burnt up all their fuel and when this happens they collapse under their own weight because they're so huge. This collapse is so violent it squashes the helium and hydrogen together with such a force that it starts an entirely unique set of nuclear reactions — producing the heavier elements like carbon and iron which make up planet Earth.

Without the collapse and subsequent explosion of these massive stars, the Universe would be empty of all the heavier elements like the metals that make our life interesting. It takes the extraordinary heat and pressure of these gigantic explosions, which we call supernovae, to seed the Universe with the components necessary for life. Supernovae explosions are so powerful they outshine billions of stars during their death throes. The picture below taken by the Hubble Space Telescope shows these leftover remnants being flung out into space billions of years after a supermassive star exploded. (This particular one is called the Crab Nebula and measures trillions of kilometres across.)

Each time these new elements from a supernova explosion are ejected out into space, gravity once again clumps them together to form the next generation of stars. But each new generation of stars is typically smaller than their parents' because the fragments have been further dispersed and less fuel is left over to build the next generation of stars. And so the process of mother and daughter stars is repeated over cycles of billions of years until we finally end up with a fourth-generation star like our own sun which has a planet like Earth circling it. It's a well-known cycle and we've seen each stage of the reproductive process with our telescopes many thousands of times. It's as well established as the life cycle of the butterfly.

We can tell what generation each star is by measuring what's inside them. It's a bit like breeding dogs. You can tell what generation a cross between a Labrador and an Alsatian is by determining how much Labrador or Alsatian is in the offspring. It turns out our own star, the sun, is a fourth-generation star, which means you and I are made up from the leftover fuel of stars that have previously exploded and re-clumped together. Without these extraordinary cooking machines, the supernovae, we wouldn't be here.

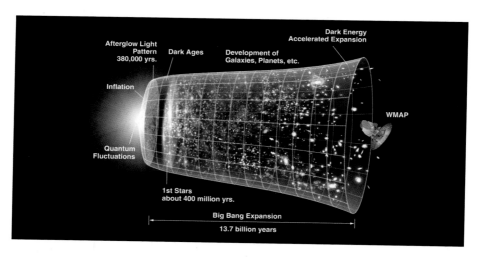

Everything fits together perfectly and matches our observations in Chapter 4, which provided overwhelming evidence for the Universe being created in a Big Bang 13.7 billion years ago. But this can only happen if a literal interpretation of Genesis is completely wrong. In Genesis the timescale is wrong and more importantly the order is wrong. The stars came first by a long way. We can be as sure of this as we can be sure of the order in which caterpillars and butterflies are produced.

But what does all this mean for our Compass of Truth? If we believe Genesis is the literal word of God then its claims must fit in the Gold region right next to all those things we think are FAPP True (For All Practical Purposes True). But as soon as we try to put Genesis in that Gold region we find there is an impossible conflict between Genesis and our beliefs about how coffee cups cool down, how police radar guns work and what we know about explosions. This conflict forces us to either call into question our understanding of coffee cups and police radar guns, or we need to reposition Genesis somewhere else on our Compass of Truth. You simply can't have contradictory items right next to each other on either side of the boundary between Absolute Truth and FAPP True.

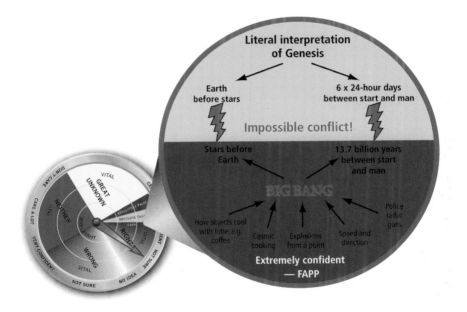

The only conclusion is the Bible cannot possibly be the inerrant and literal word of God despite the Church making this claim for some 2000 years. In the same way we must also rule out the Koran (Qur'an) as being the literal word of God as it too claims the Universe was created in a similar fashion in six days ... 'Your Lord is God Who created the heavens and the earth in six days' (Qur'an 54:7).

We therefore need to find somewhere else to put the Genesis account of creation on our Compass of Truth. This brings us to our second option, which says Genesis is simply an allegory designed to illustrate a key point or teach us a valuable lesson. In that case Genesis would now move to the Neither region where allegories reside.

OPTION 2: GENESIS IS AN ALLEGORY

Because the evidence for the Big Bang is so compelling and because it rules out a literal interpretation of the Bible, the most common position chosen by Christians is the second option, where passages in the Bible are viewed as allegories designed to illustrate a point.

An example of this stance was the recent heated debate between Fox News host Bill O'Reilly and the atheist Bill Maher. O'Reilly repeatedly said the stories in the Bible were parables and shouldn't be taken at face value nor should they be interpreted literally. Nevertheless he claimed Genesis was inspired by God who told the ancient prophets exactly what to write down so we could all learn valuable lessons from the stories contained in the Old Testament.

Allegories are difficult to deal with because they're firmly in the Neither category of our Compass of Truth. They're like works of fiction or novels. They do not claim to describe actual events or to describe Reality, but rather they try to illustrate a concept or an idea by telling us a story. As soon as we agree anything is a parable we're automatically prevented from deciding whether it's true or false. Instead we must treat parables in the same way as we would treat other works of fiction such as the novel *The Great Gatsby*. We can't, for example, question whether the representation of Jay Gatsby's life was accurate or not — for the very good reason that Gatsby never lived. It is a meaningless question. But this doesn't mean the story of *The Great Gatsby* is of no value. It may make us think carefully about our passions, our dreams or the futility of human existence. But does this mean we need to accept Bill O'Reilly's assertion that because Genesis is a parable it is then beyond reproach? Fortunately not.

The first requirement of any allegory is that we should know it *is* an allegory and it wasn't meant to represent a description of something that actually happened. In the Bible this is usually made very clear and we're normally told when a story is an allegory.

For example, in the New Testament, Jesus told the Parable of the Prodigal Son or the Parable of the Good Samaritan. In each case we (the readers) are clearly told Jesus made up the parable from common everyday events to illustrate a deeper spiritual point. Furthermore, we're left in no doubt as to what the moral of the story was. This is in direct contrast to Genesis where it's told as a 'factual' account of the creation of the Universe. This is further evidenced by the fact that for 2000 years the Christian Church always interpreted it as being a factual account. It was only when Genesis came into conflict with Science that believers began to claim it was an allegory.

The second requirement of a good allegory is that it should illustrate a moral in a way that makes the purpose of the moral either clearer or more vivid to the reader. If the story confuses the moral the author is trying to get across, we would all normally agree it was a poor allegory. So if there was a parable that said something like: 'Just as the butterfly turns into a caterpillar … so beautiful things that soar and fly will fall to the Earth and crawl', we would be rightly confused. Our experience is that butterflies don't give birth to caterpillars. In fact it is the exact opposite! It's the caterpillars that crawl on the Earth and then turn into butterflies. If we ever came across a parable like that we'd say the author was ignorant of the basics of life and the parable was a poor one.

Now if God was behind the allegory of the origin of the Universe in Genesis as Bill O'Reilly claims, why did God get his facts back to front? The allegory would have been much less misleading if the sun came first and the Earth came second. And it's hard to see what would be lost from any moral if the order was indeed correct. Surely, if I can come up with a different allegory that conveys exactly the same moral points but which more accurately reflects Reality and everyday life for a person 2000 years ago, my allegory is superior. It's not hard to do this.

We can start the story of creation by saying something along the following lines: 'In the beginning God created a great fire which spread throughout the heavens. Out of this great fire God cooked up all the ingredients of the Universe. Next God drew these ingredients together as a baker does when he forms loaves out of the dough. After a while God cooked these loaves …' and so on …

It doesn't take much imagination to create any number of allegories that a person living in ancient times would have understood and which also accurately describe the origin of the Universe. Why then did God choose such a misleading and inaccurate one? This brings us to yet another question: what exactly *is* the moral of the Genesis account? If it was there to illustrate a point, what exactly was that point? After all, Genesis is the opening account in the Bible and its message must surely be of prime importance. We can't simply claim the purpose of Genesis was to show God was our personal creator and the originator of the Universe because supplying all that additional detail about the days and the order of creation adds nothing to that claim, particularly when the detail is false.

Think of it this way. Suppose you go around to a friend's place and they proudly take you into their living room and show you a painting they've just bought, which they claim is an original Van Gogh. Initially, you believe them but express surprise at how they could afford such a masterpiece on their limited income. To allay your doubts they bring out a document of authenticity that came with the painting detailing all its history. But as you read the document you find it is full of error after error; it claims Van Gogh used latex paint (which wasn't invented until the twentieth century) and says he was living in Australia when he painted it. These false details clearly detract from the claim it's an authentic Van Gogh rather than add to it.

In the same way, if the purpose of Genesis was to show that God worked for six days and had the seventh day off, and therefore we too should do the same, it's a rather poor analogy because those days were actually billions of years of unspecified time. Surely, the Bible could have used a much better analogy.

In summary, it seems Genesis fails all the normal and reasonable requirements of a good parable. It's misleading compared with the facts and doesn't illustrate any particular moral point that followers can agree on. This leaves us with only one final option because we've eliminated all others.

OPTION 3: GENESIS IS A MADE-UP STORY

Genesis is like all other 'creation myths' and was invented by ancient humans to explain their origins and to give comfort concerning their status and purpose in life. Its final resting place is in the Red region of beliefs that, no matter how passionately held, are nevertheless false. This conclusion is devastating for the Bible and the Koran because it then undermines the authority behind *all* its moral instructions — whether on contraception, the eating of pork or homosexuality. If the Bible can't get the most important fact about humankind correct — our origin and our status — we should have little confidence the rest of its instructions are divinely inspired and *always* represent the very best way to organize society and live life.

Of course this doesn't mean the Bible never contains good instructions or that it can't be of considerable help to many people; quite the contrary. I believe it contains many exceptionally good poems and is a good accumulation of humankind's wisdom up to the point it was written. For example, the instruction 'Do unto others as you would have them do unto you', which is written in Matthew (7:12), is excellent advice, just as it was when written by Confucius six centuries earlier or when it was recorded long before in the Mahabharata (5:1517).

This instruction, known as the Golden Rule, is valuable without the need to invoke any divine authority for its authorship and it is valuable purely because it has been proven to improve the quality of any society which adopts it. Conversely, the instruction to kill anyone who does any work on Sunday, including picking up sticks or lighting a fire, is surely in error (Numbers, Deuteronomy and Exodus) just as is the Bible's assertion that women, the blind and handicapped people are second-class citizens is also wrong.

Let me finish this section by illustrating this point with an analogy. Would you trust an accountant who continually makes mistakes in her addition and whose tax advice was so confusing and so misleading that her last 1000 clients misunderstood her and did the wrong thing? Of course not! Likewise, finding Genesis belongs in the Red region of things we believe but are wrong raises questions about everything else attached to that chapter of the Bible — which means questioning the entire Bible itself.

GENESIS AND THE BIG BANG SO FAR

- The Genesis account of creation in the Bible must lie either in True, False or Neither regions of the Compass of Truth.
- But no matter how it is interpreted, the belief 'God created the world in seven days' is irreconcilable with the Big Bang.
- The Genesis account fails on three levels:
  Literal: the timescale and order don't match.
  Allegory: it's a poor one that fails the requirements of a good parable.
  Man-made: lacks wisdom in certain areas.
- We can now prove FAPP that the entire Universe was flung away from a single point 13.7 billion years ago, the Big Bang, and so the Big Bang belongs in the FAPP Green region of the Compass of Truth close to the Gold region of Absolute Truth.

CHAPTER 10
A MOMENT'S REFLECTION

I want to pause for a moment to reflect on the following stories. Each person here represents millions of others who have been harmed because of a dangerous belief that was allowed to survive in society. The stories speak for themselves and need no added commentary from me other than to say that this brings home in flesh and blood the consequences of beliefs and why we need to protect Truth.

Christian woman sentenced to death in Pakistan 'for blasphemy'

Asia Bibi, a 45-year-old mother of five, was sentenced to hang in Pakistan after being convicted of defaming the Prophet Mohammed. Her crime was allegedly saying she believed in Jesus rather than Mohammed, which is contrary to the Koran and sharia law.

Death for being homosexual

English courts forced one of the world's greatest mathematicians, Alan Turing, to be chemically castrated after being found guilty of being a homosexual, which is contrary to the Bible. He then committed suicide.

Likewise, the great writer Oscar Wilde was imprisoned for homosexuality; the experience ruined his health, eventually killing him and depriving the world of a beautiful writer and wit.

Homosexuality occurs with the same frequency in all mammals as it does in humans.

Killed for being independent

A 27-year-old Kurdish man stabbed his German-born wife in both eyes with a knife with such violence the knife broke. He then beat her with a billiard cue and finally ran over her because she was 'too independent'.

These images and stories only scratch the surface. Other examples include fatwas issued against people drawing a cartoon of Mohammed, honour killings for marrying a person of the opposite faith and the genital mutilation caused by female circumcision. Millions of people's lives have been ruined by beliefs not founded in Truth.

CHAPTER 11
SOMETHING SPOOKY IS GOING ON

If you're beginning to think I'm a 'dyed-in-the-wool scientist' who has little room for 'mystical' experiences, you'd better fasten your safety belt, because we're about to take a radical change of direction.

I'm going to tell you about an experience I had that went completely against everything I had previously believed. I'd much rather it hadn't happened, because until then my view of the Universe made comfortable sense in my own mind. Sure there were many things in that vast White region of the Unknown I didn't understand, but this threw a spanner in the entire works. Suddenly, my complacency came to a screeching halt. I'm sure I never unconsciously invented this event and I certainly had no motivation for wanting it to be true. If anything it makes this book far more difficult to write and I would have preferred it hadn't happened.

But I have to be brutally honest with you.

The problem I'm going to have as I tell you about my experience is that, for you, it will *only be a story*. You're hearing it second-hand and consequently it can't possibly have the same impact as if you were right there at the time. I know I wouldn't have believed it if someone else told me. I would've thought they were a bit of a nutcase. But I swear it's true.

THE HAIRS ON THE BACK OF MY HEAD

I'd heard the expression 'the hairs stood up on the back of my head' many times before. But when it actually happened to me, the experience was totally different to how I imagined it would be. I thought the expression was something to do with being scared or being angry because the only time I'd ever seen this happen was when cats or dogs fought. But when the hairs on my head literally bristled it was because I'd just experienced something that totally spooked me. Something so outside my world view, something so unexpected, that *everything* I'd taken for granted and accepted as true and real until that point was suddenly ripped apart.

It all started innocently enough. I was out for drinks in a bar in Parnell, Auckland called Iguaçu. I was sitting at a small table with four friends when an attractive woman with masses of long black hair came and sat down with her friends at the next table, which was very close to ours. We never spoke and I never made eye contact with her. I was simply aware of her presence because of her proximity.

On any other night she would have quickly faded from my memory as just another person in the thousands we randomly bump up against in our daily lives before passing on by. So I continued talking with my friends for another hour before it was time for me to leave and meet another friend down at the Auckland Viaduct. I left the bar, walked to my car and then drove the short 10-minute journey down to the waterfront. But in typical Auckland fashion there were no car parks so it took me another 15 minutes of circling before I could find an empty space.

A short walk took me to the restaurant where I was due to meet my friend. As I opened the door to walk in, the woman with masses of black hair who'd been sitting next to me at Iguaçu opened the door on the opposite side of the restaurant. We walked in at exactly the same time from opposite sides of the room. We both headed towards the maître d' at the same time. As we met I said, 'This is a coincidence. Aren't you the woman that was sitting next to me in Iguaçu?'

She replied, 'Yes, how very strange we both arrived at the same place at the same time.'

The thought immediately flashed through my mind that there was no way she could have known I was coming to this restaurant, and she couldn't have followed me because I'd taken such a tortuous route while looking for a park. So it really must be a coincidence. I noticed that Chris hadn't yet arrived and so I said, 'My name is Kerry. It's nice to meet you.'

She replied, 'Well, this really is a coincidence. My name is also Kerry!' With that we both sat down and ordered a coffee while waiting for our respective friends to arrive. We got on well and there was good chemistry between us and so the 15 minutes of conversation passed quickly. Before we separated to meet our friends who had finally arrived, we exchanged phone numbers and agreed to catch up again.

Unfortunately, I had to travel to the UK for work so our dinner would have to wait.

THE NIGHT AFTER

About two months later I was back in New Zealand and decided to call Kerry to see if she was still interested in dinner. We arranged to eat at the Harbourside Restaurant on the Auckland waterfront overlooking the sparkling Waitemata Harbour. Kerry was fabulous company. The food and wine were excellent and before I knew it, it was close to midnight.

At that time I was living 130 km away in Hamilton so I told Kerry I really must be getting along otherwise I was going to have a rather late night. As I was about to leave she said, 'Why don't you stay the night at my mother's place? I'm living with her while I'm in New Zealand and there's a spare bedroom you can use.' I thought about it for a moment and nearly said no, but it was a rainy winter's night and the thought of driving so far at midnight wasn't that appealing. I made a snap decision and said yes.

I slept upstairs on my own while Kerry and her mother slept downstairs on the lower floor.

So far nothing unusual had happened. But the next morning when I came down for breakfast my world flipped upside down. I said 'Hi' to Kerry and her mother as I walked into the kitchen. Kerry looked intently at me and said, 'Wow, you had an interesting flight back from America, didn't you?'

I was completely mystified by this unexpected statement and so I asked her, 'What do you mean?'

She replied, 'You were sitting on the plane when your seat lifted up and moved back five rows and came down next to a tall woman with very bright and long red hair.' She paused for a moment. 'But I don't understand the "smelly feet". What was that all about?'

The hairs on the back of my head suddenly stood up. It felt like a cold steel shaft had been plunged into my heart. I was in total shock. 'But how do you know?' I asked.

She replied, 'I see things in my dreams.'

I was absolutely gobsmacked because what she said was so accurate it couldn't possibly have been a guess. It related to something that took place on the other side of the world. Something of little consequence that I certainly hadn't told anyone about. Here's what happened.

THE FLIGHT HOME

On my way home from the UK I called in to visit Intel in San Jose for a business meeting. Afterwards I went to the airport to catch a little commuter plane down to Los Angeles where I'd then catch a 747 for the 13-hour flight back to New Zealand. As I waited to board the plane in San Jose an announcement over the loudspeaker advised us our plane was delayed because of a mechanical problem. I turned to the woman who was standing in line beside me and said, 'It doesn't give you much confidence when they tell you the plane has a mechanical problem, does it?' She was very tall, well over 6 ft, and had vividly dyed bright red hair. The plane was obviously causing the mechanics some trouble and so to pass the time we started chatting. Before long she was telling me about the internet company she'd set up in San Francisco. Thirty minutes later another announcement from the loudspeakers advised us our plane was finally ready for boarding.

We said goodbye in the way travellers do to people you'll obviously never meet again and we each went about our own business of boarding the plane. I settled down into seat 1D, ready for the short flight to LA. For some reason I turned around and glanced back down the plane. There, exactly five rows behind me, was the woman with the bright red hair. Without thinking I got up, walked back down the aisle and introduced myself to the middle-aged woman sitting next to her.

'Excuse me ma'am. I'm sorry to interrupt you. I'm sitting up in first class and I was wondering if you'd like to swap seats with me? I was talking to this woman next to you and found her most interesting and as I've got another long flight after this one I'd enjoy finishing our conversation.' Sure enough, the woman was only too keen to upgrade and in a moment I was sitting exactly five rows behind where I originally started. The conversation was interesting and the short flight passed quickly. As I left the plane I said goodbye to the woman with the bright red hair one last time. I never saw her again and thought nothing more of the entire event.

Two hours later I was upstairs in the 'bubble' on board my 747 heading back to New Zealand. But this wasn't going to be a comfortable flight. Thirty minutes after take-off the elderly woman sitting next to me took off her shoes. She'd just flown directly from London to LA and, phew, I've never smelt such fetid feet! They were just awful. But what could I do? I could hardly tell her. For the next

13 hours I constantly battled against this nauseating smell. No sooner would I start to doze off than I'd get a whiff of her feet, which would jerk me awake as effectively as smelling salts. It was a *long* flight.

Neither of these events was important to me. I certainly hadn't told anyone about them as they were just mere details in my overseas travels, but here they were, being recounted to me 12,000 km away by Kerry. By someone who had absolutely no connection with the events or anyone related to them. To say that I was stunned was an understatement. I've since thought about all the possible ways Kerry might have been able to find out about these two separate events, but I haven't been able to think of anything at all. I don't talk in my sleep and besides we were in separate parts of the house. The two women on the flights were obviously completely unconnected and were met randomly. And Kerry's description of the woman with the long red hair and the seat floating five rows back was too accurate to be a random guess.

When I questioned Kerry about her dreams she told me she hates such 'visions' happening to her. They pop into her mind at unexpected moments without her control and they don't seem to have any rhyme or reason — other than always being true. She can't do it on request and it's never of any help to her personally. As far as she knows, they started when she was hit by a motorcycle when she ran across the road as a young girl of six.

WHAT CAN ALL THIS POSSIBLY MEAN?

So this raises the question: what can it all possibly mean? To be honest I have absolutely *no* explanation at all. I could easily make up dozens of different explanations that are all compatible with these events, but this does *not* mean any of them are correct. For example, I could say Kerry is a psychic and her ability is evidence of a spiritual world, or I could say it's all to do with extra dimensions in space–time or some sort of quantum interconnected weirdness no one fully understands yet. Both explanations are perfectly compatible with my spooky experience. But it may be something completely different.

The real answer may be so far beyond my current knowledge that I can't even imagine it in my wildest fantasies. And this brings us back to the key point about that vast White region of the Great Unknown on our Compass of Truth. We *hate* seeing things in there we don't understand and so we incessantly make up explanations. Unfortunately, history tells us we get it wrong most of the time.

ZEUS AND SPARK PLUGS

When the ancient Greeks saw lightning smashing into trees and burning them to the ground they invented a God to make sense of it all. Those dangerous thunderbolts were unleashed whenever this new God called Zeus was angry at our bad behaviour and all we had to do was make some sacrifice at his altar and, sure enough, before long they stopped. It all made sense in a time when wars were fought with spears and shields. And when they heard the rumbling sound of thunder they cast around for an explanation and decided the nearest match was the rumbling of Zeus' chariot wheels across the marble floors of Heaven as he raced back and forward in his anger.

But today we no longer believe in Zeus because we've learnt how to make our own lightning bolts. Every day when we turn on the ignition switch, the spark plugs in our cars make miniature lightning bolts that power us off to work. We no longer need Zeus because we have a better understanding of the Universe. We've mastered electricity.

One by one our primitive beliefs have fallen prey to the relentless march of Science, which has almost always revealed a universe that is more bizarre than we could ever have imagined — but one which is also far more interesting.

This invention of Zeus provided an explanation for events they didn't understand, and having any explanation makes our brains happier. We can then relax and get on with living without that nagging doubt of uncertainty hanging over us. That's why we invented stories such as Noah's Flood to account for rainbows, or possession by demons to explain epileptic fits. History is littered with a huge range of bizarre beliefs and explanations we've invented to make sense of and describe the Universe around us. In most cases we've relied on our normal daily experiences and so our God is a wise heavenly father or a great warrior, depending on the culture we were brought up in.

Our brains are wired with a built-in need to spot patterns and understand the Universe. The reason for this is if we know how the Universe works, we can act accordingly. By constantly looking for patterns and order we notice the seasons come and go in regular cycles and this allows us to plant and harvest our crops at the right times. Understanding patterns and finding reasons for things allows us to conquer the Universe instead of being at its mercy. And so any brain that has this feature of finding patterns or making explanations built into it will have a natural advantage over brains without the feature. It's not surprising, then, that we're hard-wired to search out patterns lurking unsuspectingly behind events.

The problem is, for most of our existence we haven't had anywhere near enough knowledge or information to explain most of the things we observed. But that didn't stop us; we simply made up a host of explanations that seemed reasonable to us at the time.

LIMBIC MASTERS DON'T ALWAYS NEED EXPLANATIONS

Until we discovered electrons, the idea that microscopic particles millions of times smaller than anything our eyes could see were the cause of lightning would have seemed absurd. To the ancient Greeks, Zeus was probably a much more sensible solution than the correct explanation.

People mistakenly think that admitting we don't understand something is a sign of intellectual weakness, a lack of knowledge or an admission the rest of our beliefs are somehow flawed or limited. But the opposite is actually true. One of the surest signs a person has risen above their primitive natural self and has now become a Limbic Master (someone who has mastery over their primitive limbic brain) is that they no longer need to 'blindly invent' explanations for things they don't comprehend. Limbic Masters are able to say, 'I don't understand this event. At present it's beyond humanity's knowledge just as lightning was beyond the ancient Greeks' knowledge. I don't need to invent an explanation. No explanation is a better hypothesis than a bad explanation.'

Admitting we don't know the answer to one particular question doesn't mean we need to call into doubt everything else we know. We understood how chemistry produced fires and explosions and those descriptions were 100 per cent correct even if we didn't yet know about nuclear explosions or how the sun was powered. This is an important part of the power of realizing our Compass of Truth has multiple regions. We can be quite comfortable with saying there *must* be things in the Great Unknown White region and those things don't undermine the validity of the things we place in the Gold region of Absolute Truth or in the Green region of things we believe For All Practical Purposes to be true.

We therefore need to be extremely careful when claiming our lack of understanding of certain physical events is somehow evidence of the supernatural, because history tells us we're usually mistaken. One by one the superstitions we used to hold so dear have fallen by the wayside as our sphere of knowledge in region three (things we believe and are right about) has gradually widened. But this true story about Kerry is also a reminder to be more humble in our claims that our scientific knowledge has all the answers.

For example, physicists began to put forward a so-called 'Theory of Everything' in the 1980s which many scientists thought would soon provide a single mathematical equation that explains all the forces and particles of nature. They still haven't managed to do that yet, but even if they could, it would fall far short of answering some of the most profound questions concerning our lives.

As we'll see in Chapter 16 when we examine our 'soul', any claims about a Theory of Everything are wildly extravagant because such a theory wouldn't even know how to *begin* answering the question of consciousness.

CHAPTER 12
THE PSYCHOLOGY OF BELIEF

MY RELIGION WORKS FOR ME

One of the most common arguments used by people to 'prove' their religion is true is to say something along the following lines:

'It doesn't matter what you say, I know *my* God is *real* because I have a personal relationship with him. Regardless of any so-called "facts" to the contrary, I experience God's presence and power directly which is why I believe and have faith.'

At first glance this seems like an impossible argument to overturn because no one is in any position to question someone else's personal experience. We have no right to say what goes on inside their heads or to question their Reality.

Or do we?

Many years ago I remember watching a boxing match on TV where both fighters went to their corners after receiving their final instructions from the referee. Each fighter then got down on his knees before the bell sounded and prayed to his God for victory — one prayed to Allah and the other to Jesus. I don't remember who won the fight or which 'God' helped his boxer to victory, but it brought home to me the stark reality that millions of people passionately believe in a God that can't possibly exist.

Both boxers passionately believed in their God, but the problem was, these two Gods are completely incompatible according to the view most commonly held by their followers. Both Gods couldn't have been the single *unique* creator of the Universe any more than you can have two mothers who both gave birth to you. If two women claim to be your biological mother, at least one of them must be wrong. Likewise, both Gods can't be the *only* way to salvation just as you can't have two winners in the boxing ring at the same time.

Of course, this doesn't rule out the possibility of there being many gods (spelt with a little 'g'), such as those in Greek mythology, who don't make the big claim of being in sole charge of your destiny or of the running of the Universe. These little semi-gods could easily coexist just as fairies and elves at the bottom of your garden might. What we're focusing on here is the unique 'capital G' God who is solely in charge of the Universe. But let's be very clear, in this chapter we're not claiming there is *no* God — there might well be one — what we're saying is there can't be *two* of them. And because there have been thousands of such 'capital G' Gods in history then millions of people must have had a satisfying personal relationship with a God that didn't actually exist.

This brings us to the central questions of this chapter:

- How can a God that cannot possibly exist seem so real to so many people?
- What actually makes these religions 'work'?
- Can we use the knowledge of how these beliefs work to make our own lives better?

Maybe a good place to start is to find somewhere else where there's a total disconnect between the **subjective Reality** experienced by individuals and the **actual Reality** of how the Universe works.

Medicine provides us with a perfect example.

WHAT IS THE MOST EFFECTIVE DRUG IN THE WORLD?

Ask yourself the following question: 'What is the most effective drug in the world?' Or to put it another way, 'Which drug has cured or helped more people recover from a wider range of illnesses than any other drug?'

Without doubt the most effective drug in the world is also the only one that has absolutely zero active ingredients — the **placebo**. Whether it's for cancer, heart disease, skin irritations, depression or any medical complaint you can name, placebos produce statistically significant improvements in patients every day of the year throughout the entire world. It's probably just as well placebos are so powerful because for centuries that was the only thing doctors had. We now know that most pills and potions before 1900 either had zero therapeutic value or were actually harmful to their patients (e.g. blood-letting).

Yet despite having no genuinely effective medicine, doctors continued to be held in extremely high regard because their potions worked anyway. It was all down to the placebo effect. As long as you believed the pill was going to work — it did! Unfortunately, 'quack' medicine hasn't completely ceased in modern times and there are still many popular treatments such as homeopathy which rely entirely on the placebo effect.

Because totally inactive placebos are so powerful the only way we can determine if any drug really works is by comparing it with a placebo in a procedure called the 'double blind' trial. If a new drug can't outperform a placebo under these controlled conditions, it isn't licensed. What's particularly interesting about placebos is that patients not 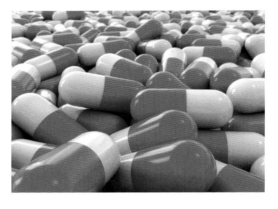 only subjectively feel they are getting better, but they actually do get physically better. Their cholesterol is lowered, rashes disappear and they recover from their illnesses faster with a placebo than without one. This is quite an extraordinary result. We might even say a placebo is more 'real' than some religious beliefs because it produces measurable physical changes to a person in addition to the subjective psychological feelings they experience inside their heads.

What the placebo demonstrates is the incredible power of having a **belief in something,** even if that something is completely and utterly non-existent. Psychologists have studied this phenomenon for many years and have come to the conclusion that belief itself is the magic ingredient and it's less important what you actually believe in. All you need to do is believe in something and this act of faith will make a profound difference to your life. It doesn't matter if it's a rabbit's foot, magic water or some mythical God.

While this is part of the reason a religion can work even without a God it is by no means the end of the story.

SELF-HELP BOOKS

Another insight can be gained by considering self-help books. If you take 1000 self-help books and list the five key points from each book and then add up the most common 'secrets to success', what do you think would be the Number 1 most popular rule?

The answer is 'Believe in x' ... where x takes a bewildering variety of forms.

It could be to believe in yourself, or in the Universe to provide, or in your destiny, your ability or in God. Once again, it seems having a belief is incredibly powerful regardless of what you actually believe in. This brings us to the question of how can a belief work all on its own without something solid underneath backing it up?

Unfortunately, the English language lets us down when we try to answer this question because the word 'belief' has two different meanings:

1. Belief: confidence, trust or hope in something, e.g. Agassi had belief in his tennis ability.
2. Belief: the set of ideas you think is true, e.g. I believe the Earth is round.

Let's start with the first definition.

BELIEF = CONFIDENCE: PERSISTENCE AND AN UNSHAKEABLE BELIEF

Having an **unshakeable belief** is like possessing a powerful engine that drives you forward no matter how many times you get knocked down, beaten or disappointed. An unshakeable belief keeps you going because if you truly believe, then it's only a matter of time before you're successful. You're convinced the outcome is inevitable; after all, that's the definition of an unshakeable belief, and so any failure just motivates you to try even harder next time. This now brings into play a powerful law of life: the harder and more often you persevere the more likely you are to succeed. Additionally, all this practice is bound to improve

your skill. Before long you'll be in a 'self-sustaining positive feedback loop' of the
type depicted below.

Reinforce and strengthen belief

Your persistence inevitably increases both your skill and your opportunity for
success and any small success you do get only serves to increase your confidence
and make you try even harder. And so a cycle of positive reinforcement has
begun.

There are plenty of examples showing how this feedback process can produce
truly life-changing and permanent results in people. For example, in one notable
experiment, teachers were told a small number of the children coming into their
new class were either above or below average intelligence.

What was important about this experiment was that none of the children given
these 'labels' were actually above or below average. They had simply been
picked out at random from the rest of the class. But when all the children were
independently tested at the end of the year by a second teacher who didn't know
about those artificial 'labels', the researchers found those who were given the
'above average' labels did better than the rest of the class and those who were
given 'below average' labels did worse than the rest of the class. The teacher's
expectation of the children had subtly changed the children's actual performances.

So it doesn't really matter if you believe in a magic ring, your destiny, the cosmos
or what your coach told you — all that matters is you keep on trying and have
a positive frame of mind. And belief is what ensures this happens. Now this is
where things get interesting. If you think about it, Religion must be the most
powerful system ever invented for instilling belief and hope because if you believe
in a God who made the *entire Universe* then surely he can make things work out
for you. No matter what happens, it must all be part of God's plan.

As the Bible says, 'Things will always work out for the best in the long run if you
love God.' In this way a belief in God gives you the ultimate level of confidence

possible, which in turn means you're more likely to be happy and succeed — whether or not God exists.

THE SELF-FULFILLING PROPHECY

Another outcome of belief is something called the self-fulfilling prophecy. Here's how it works. Suppose a man goes to a clairvoyant and after looking into her crystal ball she says to him: 'I see you are going to marry someone in a uniform.' The instant she says this, his chances of actually marrying someone in a uniform increase by 1000-fold for the following reasons.

Firstly, he's keen to meet his soul mate and marry her — after all, who isn't? Secondly, he believes in the clairvoyant, otherwise he wouldn't have gone to see her in the first place. The net result is that he switches his 'radar' on to full alert for any woman in uniform. It could be a nurse, a flight attendant, a supermarket checkout woman, a policewoman or any one of hundreds of other careers. Suddenly, the world is filled with possibilities.

It's a bit like when you buy a new car. Everywhere you look you see the same model everywhere you go. And why? Because unconsciously you're paying attention to them. Likewise, as soon as you see a woman wearing a uniform you unconsciously give her that extra little bit of attention. Without the 'priming' you got from the clairvoyant you might have dismissed her as 'out of your reach' or conversely 'not good enough' for you. But because you know one of these women in uniform is going to be your soul mate you at least give it a try. So you pluck up your courage and walk over to her because you'd be mad to let your potential soul mate walk on by. This act opens up a possibility and increases your chances of success.

Conversely, if you come across someone else who never wears a uniform you'll unconsciously find a fault with her because there must be some reason why she's not going to be your soul mate. So you walk on by, closing off that opportunity, which reduces your chance of marrying a woman without a uniform.

It turns out it wouldn't really matter what the clairvoyant said to you. She could have told you your soul mate was tall, or had blonde hair or came from a foreign country. Whatever she said, as long as you believe in the clairvoyant, it's much more likely to happen.

So where does this leave us?

NAKED BELIEF

We've seen many things like 'good luck charms', placebos and non-existent Gods — all of which have zero actual power on their own — give people a belief to carry on with a positive mental attitude. This belief works in spite of and independently of these mystical items. This therefore raises an interesting question: rather than believing in some empty concept like a false God or a charm, why can't we just believe in the psychological process of belief itself?

I do and I've found it works brilliantly.

Many times when I'm completely beaten and have nothing left to hold on to, I say to myself: 'Just keep on going and you'll be successful.' I hold on to naked belief itself, because there is nothing supporting my belief: no God or destiny, and in many cases not even any logical reason for my belief. So I keep on going and try to work smarter and harder. Sure, it doesn't always work immediately and magic hardly ever happens, but over the long haul it's been a very successful formula. For more on this you can read my No. 1 bestselling book *The Winner's Bible* where I explain with practical examples how you can get an unshakeable belief and use this to your advantage.

At this stage I feel it's important to draw a distinction between the sort of naked belief I am talking about here and the sort of belief referred to in the book *The Secret*. In that book your belief somehow attracts what you want to you because there is some cosmic mechanism that swings into place to provide your desires. All you have to do is believe strongly enough and it will be provided to you by some mystical agency. I am absolutely convinced, and all the evidence seems to support me, that belief works because of the positive feedback loop in the diagram above and not because of some external agency which provides things to you.

So my challenge to you is to see if you can believe in the strength of naked belief without the need for a crutch or item like a 'rabbit's foot' to give you power. I've personally found it's very liberating to experience raw, naked belief all on its own.

BELIEF AS A MODEL OF REALITY

Earlier in this chapter I said there was a second type of 'belief' which is concerned with our internal model of Reality. To see how this works, imagine you're sitting in a train reading the morning's newspaper while you wait for your train to leave the station. You look up from your newspaper at the passengers in the train parked next to yours and wonder where they're heading and what they're up to.

A few minutes later your train begins to pull away from the station and slowly the passengers in the next carriage begin to slide away behind you. Then, all of a sudden you reach the end of the neighbouring train only to discover the other train was moving and you were stationary all the time. For a millisecond your mind has a minor spasm as it recalibrates and accepts you weren't moving at all. This common experience illustrates a vital point: we don't experience the raw inputs of nature directly but rather we interpret them according to our internal model of Reality. In this case our model of Reality was that if everything outside us appears to be moving, it probably means *we* are moving.

Our internal model doesn't just interpret physical sensations such as moving images; it's also used to interpret complex things such as human behaviour. This explains why two people can view the same statement someone makes in a completely different way. One person might think Obama's healthcare plan is a sign of his genuine concern for the underprivileged while another person might think it's a sign of his 'big government anti-free market left-wing bias'. It all depends on your internal model of who you think Obama is and what drives him. We can represent this state of affairs schematically in the following diagram.

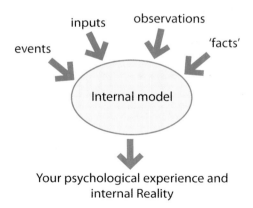

The reason we have an internal model is that it allows us to drastically simplify the Universe so we can deal with everyday life rather than being overwhelmed by zillions upon zillions of apparently random events. We notice lightning and thunder always seem to go together and so after a while we work out they're not two entirely separate things but are merely different manifestations of the same electrical discharge. Or we discover 'gravity' and see it explains a whole host of things that previously didn't make sense. Now we know why apples fall to the ground, why the planets circle our sun and that in turn leads to an understanding of why we have a 24-hour day. In short, we find connections and patterns in nature that both simplify and explain the Universe. This makes the Universe far less scary and easier to deal with.

Finding connections and patterns like this turns out to have another advantage for us because it also allows us to predict what's going to happen next. We observe that small animals in nature generally produce quiet sounds and big animals generally produce loud sounds, so when we hear a sudden roar behind us we jump just in case it's a tiger about to eat us for lunch. It's therefore not surprising our brains have evolved to be extremely good at finding patterns and connections between things and then constructing some sort of model that explains them. This pattern-finding ability is so good we can even find patterns where none exist, such as when we see seahorses chasing elephants in cloud formations.

Unfortunately, there are sinister side-effects of this strong desire to categorize everything and find patterns. One such negative effect is where we pre-judge people on the basis of shared traits, such as happens in racial prejudice or in discrimination against women.

Our internal model of the Universe also includes much more abstract things such as whether there is a 'God who created us' or whether we 'evolved by Evolution through Natural Selection'. And it's when we get to these more abstract parts of our 'belief model' that things get rather interesting.

COMPETING BELIEF MODELS

You've probably seen a number of illusions where a single image can be interpreted in two completely different ways. One of the oldest of these is the face–vase illusion where you can either see two people painted in black facing each other or else you'll see a solid white vase in the centre of the picture. Both are equally valid interpretations of the image.

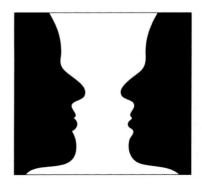

The same sort of conflict can also arise with our beliefs because it's quite possible for two completely incompatible models to fit the known facts equally well, even though at least one of them must be wrong. For example, consider an ancient person watching the sun rising each morning in the east and setting each evening in the west. There could be two explanations for this observation: either the sun circles around the Earth each day or the Earth rotates on its own axis while the sun stands 'still'.

As far as ancient people knew either model would perfectly fit the facts. Indeed it probably made more sense to them if the sun moved and the Earth stayed still because they didn't feel like they were spinning around getting dizzy. But of course we all know the story didn't end there. When we finally got around to looking at all the other stars we suddenly realized they'd have to move in the most impossibly complex motion if the Earth was indeed standing still. But if we changed our model so the Earth moved around the sun suddenly everything made perfect sense. This new model also explained why we had seasons and the length of the days varied in winter and summer.

What's important to notice here is that our 'natural' model which initially best fitted the facts — namely the Earth standing still — ultimately turned out to be wrong. In the same way, religions provide explanations which while sensible and logical 2000 years ago are no longer able to fit the facts now available to us.

COGNITIVE DISSONANCE AND MODEL PROTECTION

The possibility there might be more than one 'internal model' that can fit the facts presents us with a huge problem. If we have a choice of models to use, we won't know how to interpret an event or information presented to us. We'd end up jumping back and forth between models and because each model tells us to do different things we'd be completely paralysed by indecision. This isn't a luxury we could afford because while we were jumping between competing models deciding how to interpret that loud noise behind us we'd get eaten by the tiger. The simplest way around this dilemma is for us to settle on one 'favoured model' in our mind which we automatically use to interpret everything in a quick and efficient manner.

To ensure this model is robust our brains contain a special mechanism that automatically tries to remove any apparent conflicts between 'facts' and our model. One way it does this is by distorting the information coming in and filtering it so the 'facts always fit with our model' — sort of like wearing rose-tinted glasses.

Another way is to ignore or discount any information that conflicts with our model. This process of reducing conflicts with our model is called dissonance reduction by psychologists and is an extremely powerful part of our nature. It's part of the reason why spouses continue to believe in the innocence of their partner even when there's overwhelming evidence against them. Their internal model portrays their partner as being a kind, loving person, which doesn't fit the facts, and so one way of reducing the conflict is to discount the evidence. Maybe the police tampered with the evidence or their partner was framed by some other cunning criminal. It's also why other people can believe things that seem just plain stupid to you and vice versa.

We're all unconsciously filtering the Universe. What's really interesting about this protective mechanism is that it has to operate without us ever knowing what it's doing, otherwise we'd be alerted to the fact there was a conflict going on and once more we'd be paralysed by indecision. Just as our body has a vast range of techniques at its disposal to keep our body temperature constant regardless of the external environment, such as shivering when we're cold or sweating when we're hot, so too our brain uses all sorts of tricks to automatically protect our internal model from any conflict.

Religions have an added method of reducing dissonance because they place a huge premium on 'faith', on believing in spite of the evidence. The more 'faith' or 'belief' you have in your God the more righteous and spiritual you are. This provides the believer with the ultimate protection from Reality. Very little can get through a shield like that.

DAMASCENE CONVERSION

However, from time to time the disagreement between our model and the 'facts' becomes so great that our hidden filter can no longer cope. When that happens the only way out is for us to completely overhaul our model. This is usually accompanied by a major psychological experience like the 'judder' we experienced when we suddenly realized our train wasn't moving at the station. That's not surprising since the old inputs we've been familiar with all our lives are now interpreted differently. From then on we behave differently and become subtly different people because any alteration to our internal model also reprograms our psyche. If this happens in a religious setting (as in the story of the conversion of Paul the Apostle on the road to Damascus) we say a person had a 'conversion experience'.

The same process can also work the other way when a person suddenly realizes the God they previously believed in doesn't actually exist.

While this process of accepting that your previous world view might be wrong can be intensely uncomfortable, it is also the mark of a wise person. Immature people hold on desperately to their cherished views while a wise person says 'all facts are up for review'. It can even be a positive experience to take off your own 'tinted glasses' and try on another pair for a while to see how they fit. You might find it's as refreshing as taking a holiday.

CHAPTER 13
EVOLUTION: A *CSI* CRIME SCENE INVESTIGATION

COMMON OBJECTIONS TO THE THEORY OF EVOLUTION

If ever there was a hung jury in the court of public opinion it would have to be over whether Evolution is the true explanation for the origin of humankind. Were we personally created by a God or did we evolve from bacteria through a process of Natural Selection and random mutation? Clearly, the status of humankind, of *you*, is quite different depending on which answer is correct.

Recent polls in both the USA and the UK found the population is bitterly divided with approximately one-half thinking Evolution was the cause of humankind and the other half thinking Evolution wasn't. The disagreement between both sides is ferocious and you only have to look at any of the thousands of blogs and websites devoted to this question to see each side claiming the other is ignorant and lacking in common sense.

Because we are passionately interested in the Truth we'll need to work out where human Evolution fits on our Compass of Truth. Should it be placed with high confidence close to FAPP at one o'clock or should it be near three o'clock because we just don't have enough information to decide? Or should it be put in the Red region of beliefs that are wrong?

Having been brought up as a devout Christian in my childhood I've had a particularly keen interest in this debate. In my experience, regardless of all technical arguments people initially talk about when the debate starts, the main objections to Evolution seem to fall into five main categories:

1. The argument over design and probability

How can something as complex as a human realistically be produced by Evolution? If it takes thousands of engineers millions of human-years to design a car, surely some guiding designer must have been involved in crafting our vastly more complex and intricate construction.

2. Mutation versus transmutation

While we routinely observe animals adapting to suit their environment over generations (mutation inside a species), it seems an altogether different question for a human to have evolved from a completely different species like a whale (transmutation).

3. God-directed Evolution

Suppose Evolution actually happened and humans did evolve from earlier hominids (great apes), this still doesn't rule out the possibility that God was in charge of the process directing each step. Maybe God controlled each step of the evolutionary process?

4. Who started it all off?

Even if Evolution works, who or what created the Universe in the first place? Evolution still needs something there in the first place to work on.

5. Meaning in life

How can we have a purpose or a meaning in life if we were produced by blind Random Chance and Natural Selection?

CSI: CRIME SCENE INVESTIGATION

At the very least we'll need to answer each of the five points above in a totally convincing way. To do this I'm going to adopt a completely different approach to that normally used by most people because I don't just want to convince your 'mind', I want to convince your deepest gut emotions as well. This is necessary because I've often heard people say, 'Regardless of all the evidence for Evolution it just doesn't feel right deep in my bones. I can't put my finger on it, but emotionally I simply can't accept I'm a product of randomness.'

The approach I'm going to use to answer the first three points is the same as used by the detectives in the TV series *CSI: Crime Scene Investigation*. Points 4 to 6 will be addressed in their own separate chapters (Chapter 14: What Caused it All? and Chapter 18: Purpose and Meaning).

THE FINGERPRINTS OF GAUSS

One of the first things any detective does at a crime scene is try to find evidence that identifies who was at the scene when the crime happened. Finding a suspect's bloody fingerprints all over the murder weapon would be extremely strong evidence linking them to the crime and the ideal sort of evidence they're after. In the case of Evolution and the origin of humankind we'll begin by assuming our prime suspect is Random Chance. You'll notice I've spelt Random Chance with capitals just as you would a person's name because we're going to treat Random Chance as we would any real suspect in a court case.

You'll also see that Random Chance has a first and a second name because we want to be quite clear this is not just any ordinary type of 'Chance' such as you get in 'coincidences'. This type of chance is one which is completely and utterly without any pattern, coincidence, order or structure. It is *random*, so Chance has a specific first name. Next we need to find some unmistakeable fingerprint that's possessed by Random Chance and by no one else or nothing else — something totally and utterly unique to Random Chance. If we can do that we'll then need to see if Random Chance has indeed left its fingerprints all over our origin and whether there were any other fingerprints.

We're thus faced with answering questions in a two-stage process:

1. Is there a fingerprint that is absolutely unique to Random Chance?
2. Do we find those exact fingerprints all over our origin?

Fortunately, it turns out there *is* one fingerprint which is unique to Random Chance and no one else.

This fingerprint is known as the Gaussian distribution and is sometimes referred to as the 'Bell Curve' because it has a bell shape in outline. Although most of us came across it in high school, not many people really understand where it came from or why it's so important. Let's see if I can make it simple because you'll want to know about this fingerprint because it has left its unmistakeable fingerprint on your looks, your intelligence and even your eyesight.

THE COIN TOSS COMPETITION

The easiest way to get to know our Gaussian distribution curve is to begin by getting a group of 1000 people to toss a coin 10 times each. We then get them to report how many heads they threw in their 10 trials.

Because they're equally likely to throw a head as a tail most people will get around 5 heads and 5 tails. But some people will throw 6 heads and 4 tails, others will throw 7 heads and 3 tails and so on. There might even be some who throw 10 heads in a row.

If we do this experiment and plot the results we get the following results:

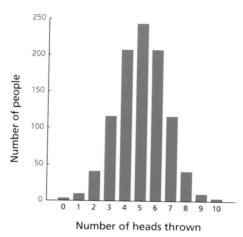

1000 people tossing a coin 10 times

As expected very few people get either 10 heads or 10 tails in a row and most score 5 of each as shown in the bar graph. And if we repeat this experiment we always find we get the same pattern. Even though each individual coin toss for every person was totally random we always get the same distinctive shape when everyone's scores are tallied.

Now let's repeat the test but instead of using a coin which only has two outcomes, head or tail, we'll use a 'spinner' with numbers marked from 0.0 to 1.0 around the outside. In my diagram I've only got 10 numbers (0.1, 0.2, 0.3 ...) around the outside because I can't show them all, but let's assume there are thousands of numbers filling all the way around (0.0001, 0.0002, 0.0003 and so on).

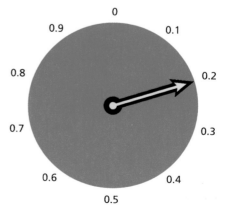

Instead of tossing a coin we ask each person to spin the pointer 10 times and write down the number beside where the pointer stops and then add up the numbers.

On my 10 turns I got:

$$0.34 + 0.18 + 0.29 + 0.85 + 0.36 + 0.77 + 0.95 + 0.11 + 0.01 + 0.58 = 4.44$$

Score after 10 spins

Once again, If I get 1000 people to spin their dials and plot their scores I get the following graph. Instead of only getting 1 or 2 up to 10 heads as I did with the coin toss, I can now get any number from 0 to 10 including all the decimals in between. That's why it's now a nice curve instead of the discrete bars we got before.

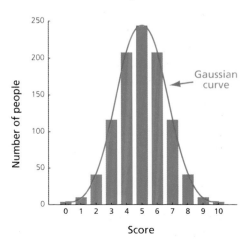

Coin toss and spinner overlaid

But just as the coin had an equal chance of being a head or a tail, so too my spinner has an equal chance of stopping above 0.5 as it has of stopping below 0.5, so it's not surprising the red curve for the spinner and the blue bars for the coin have the same shape.

Now the interesting thing about the Gaussian curve is that its shape is *extremely* well defined. It's not just any bell curve but the shape and height at each point on the red line is even more carefully aligned than any whorl of a fingerprint. For example, on the diagram I have drawn another 'bell curve' in blue but this one has the wrong shape. For a curve to be Gaussian we need precisely the right number of people to get a score of 2 and precisely the right number of people to get a score of 3. In fact we need to get precisely the right number of people to get *every* score from zero all the way to 10.

The shape is absolutely precise

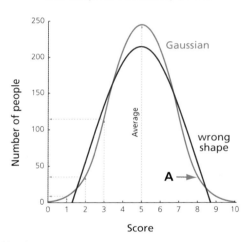

On my diagram you'll see I've marked a place with the letter A. At that point the graph must have precisely the right height and precisely the right slope. This requirement is true for *all* points on the red line.

To be a true Gaussian curve is far more demanding than any fingerprint presented in court needs to be. And we *only* get a Gaussian curve when we have Random Chance producing results like we do with our spinner or our coin. Nothing else ever gives us this shape. We can actually *prove* mathematically that a Gaussian curve is only produced by Random Chance — although we won't do that here. Thus Random Chance being the only cause of Gaussian curves is something that lies very nicely in our Gold region of Absolute Truth.

OK, so we know the shape of a Gaussian curve and we know it is produced only by Random Chance. Where do we find it in nature?

YOUR LOOKS, YOUR IQ, YOUR EYESIGHT AND RANDOM CHANCE

What do you think happens if you take 1000 newborn babies and test their eyesight?

You guessed it, most babies have close to perfect vision but some are short-sighted and some are long-sighted. And just like our coin toss, the curve representing the number of people who threw 7 or 8 or 9 or 10 heads in a row has exactly the shape as the number of babies who have increasingly long sight. We get precisely the same-shaped curve as we would get if we took all the babies in the world and simply spun a pointer around a number of times and then gave each baby the amount of long-sightedness or short-sightedness we got by adding up their pointer scores.

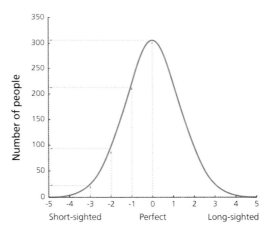

We can try fitting other bell-shaped curves to our eyesight measurements, but the only one that fits correctly is the one produced by Random Chance.

If we now test everyone's IQ we find the same curve for their intelligence with the same precise fit at every point along the Gaussian graph.

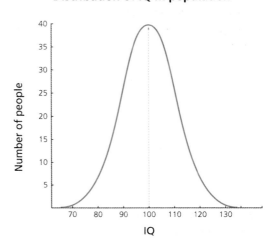

I could go on with many other human traits but I'd like to mention one last one that's particularly important — the age at which people die. You can probably guess I'm going to tell you that it's also a nice fit with the Gaussian curve, although different populations have their curves centred at different ages depending on whether they smoke or which country they live in.

So what do all these Gaussian curves mean? They mean you can't possibly have been personally handcrafted and designed by some master creator or designer. All your most important aspects, whether it's your intelligence or your good looks, are simply down to the random mixture of genes you inherited from your parents. And when it comes to that most important aspect of your entire life, how long you live, this too is simply the result of a myriad random events (genes, environment and car accidents) all mixing together to produce the Gaussian curve.

Think about it from another angle; if every aspect of a person's life was designed, surely the only conclusion we can come to is the designer wasn't very nice. After all, who would deliberately give a young child a cleft palate as their birth present? And if God wanted those children born with a cleft palate, we'd be committing an evil act by opposing his will if we used modern surgical techniques to correct the fault. The argument is no different to that used by the Catholic Church to oppose contraception.

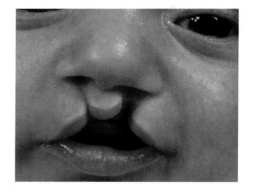

The underlying randomness we observe cuts deep at the heart of whether there could possibly have been a predetermined destiny for you or whether it has more to do with the random throw of a coin. Of course this doesn't mean everything is random in your life. You can still raise yourself above your 'natural' level by having excellent life skills and you can easily squander your native potential by poor habits. Nadal may have a natural aptitude for tennis but without practice and dedication he couldn't possibly have become a champion. That's where our individual responsibility and our own self-worth through effort comes in (see *The Winner's Bible*).

Now, just because Random Chance has shaped and left her 'fingerprints' all over the construction of every individual living today, this doesn't necessarily mean it has anything to do with the origin of humankind as a whole. For that we need to dig a little deeper.

DOES IT HAVE THE ABILITY — IS EVOLUTION STRONG ENOUGH?

The next thing any detective must do is determine whether the suspect had the ability, the skill and capacity, to commit the crime. To see how this works let's suppose our victim is a 120 kg bodybuilder who was found dead on the floor of his home, covered in his own blood. The crime scene investigation revealed the

lock on the back door had been picked open and the victim had been strangled to death after a violent struggle. It is clear the blood all over the floor and walls had come from cuts to the victim's mouth and nose after receiving many vicious punches that knocked him out before he was finally strangled. Now even though a 90-year-old woman with arthritis may have left

her bloody fingerprints on the victim it's most unlikely she was the person who beat and strangled the 120 kg bodybuilder to death. It's much more likely her fingerprints were left at the scene when she tried to resuscitate the victim after the murder had already taken place and the criminal had fled.

This hypothesis is supported by the fact that the frail elderly woman probably didn't have the strength to overpower the bodybuilder nor the skills to pick the lock with her arthritic fingers. In short, she didn't have the strength or the ability to commit the crime on her own. Someone else must have been involved.

This question of 'capacity' — of the ability to commit the crime — is the one most commonly levelled against Evolution. It just doesn't seem right that Random Chance could have pulled off the incredibly complex systems that comprise even the simplest of animals let alone something as amazing as a walking, talking, music-composing human. After all, the best scientists can't even produce a robot that can play a poor game of tennis or clean the living room and put away children's toys.

Surely, if we're stumped with all our collective genius and the years of research and development we've invested in robots, how could Random Chance have any possibility of creating a much better product? It just doesn't seem sensible does it? This issue of ability is therefore at the heart of Question 1 at the start of this chapter: *How can something as complex as a human being realistically be produced by Evolution? If it takes thousands of engineers millions of human-years to design a car, surely some guiding designer must have been involved in crafting our vastly more complex and intricate construction?*

Like all court cases, it's up to the prosecution to show beyond reasonable doubt the suspect had the ability to commit the crime. This is usually done with a convincing demonstration in the courtroom where each step of the crime is carefully laid out. So why don't we take a very complex object that seems unlikely to have arisen by Random Chance and then see if it was sufficient after all.

How about we take a car?

A car is surely an excellent example because Toyota, Jaguar, Ford and Ferrari — to name but a few — have tens of thousands of engineers working feverishly day and night to design a better car. Despite all these brilliant engineers and their computers and wind tunnels it still took 60 years to go from a Model T Ford to an E-Type Jaguar. That's clearly a huge amount of creative genius and design. Surely, we couldn't achieve that by chance, could we?

Let's try!

EVOLUTION OF A CAR BY RANDOM CHANCE

What I'm going to do is very simple. I'm going to start off with a car which is shaped like a Model T and then I'm going to let it evolve all on its own. I'll do this in the following fashion.

Step 1: Describe each car by its genes

First of all I'm going to describe the height, length and angle of the car's key components, such as its roof, hood and trunk with numbers. This is not unreasonable as the length and angle of each body part determines the styling of the car. We can think of these numbers as the 'genes' of the car. They're like the numbers on a blueprint drawing that tells the craftsmen how to fabricate and assemble each component. The picture below shows the idea with just a few of the numbers shown for our original Model T.

Clearly, we're going to need quite a few numbers to give a realistic shape for our Model T. Nevertheless, identical-shaped cars will have identical genes (or numbers) while different-shaped cars will have their own unique set of genes — just like humans.

Step 2: Breed the cars

Next I'm going to breed my Model T and allow it to have thousands of Model T offspring. The very first time we do this I'll just make identical copies of our original Model T, but for all subsequent times through this process we'll breed them in the normal way where half the genes (or styling numbers) come from one car and half the genes come from another car. In other words, we'll take two cars and combine their genes to make a third in the same way animals do with sex.

Step 3: Introduce some random errors into our breeding programme

Of course, if we just copy our original car, we'll end up with an endless number of identical cars. So what I'm going to do next is introduce a small amount of randomness into the process. In the process of copying half the genes from one car and half the genes from another car, I'll accidentally make a small mistake on a very small number of genes. It'll be like making a mistake when you copy a poem from one book to another or when you don't correctly hear what someone said. But in more than 99 per cent of the time I'll copy the genes perfectly. I'll only make a small random mistake in a small randomly chosen number (gene).

Notice we've used the word 'random' twice in the previous sentence. I'll have no choice as to which genes I'll alter, whether it was a number for the height of the roof or a number for the length of the hood, and I'll also have no choice in whether I make those numbers bigger or smaller or whether the changes will be small or great. Remember, in most cases I won't make any errors and any that I do make will all be down to pure Random Chance.

Step 4: Kill off the old cars

All this breeding will produce lots of cars so I'll kill off any car that has bred five times to make sure we don't have impossibly old cars driving around.

And that's it. Only four steps!

So, how am I going to accomplish these four steps? Well, I don't actually want to build millions of cars and kill them off because that would be far too costly and time-consuming, so what I'll do is write a small computer program that simulates this process instead.

So how do we get on when we do this?

When you run the program it's a *total disaster!* You quickly end up with cars that look weird and would clearly be completely undriveable. Some of them don't even have room for the driver. Clearly, unaided Random Chance, like the elderly woman with arthritis, doesn't have the capability. On its own Random Chance is simply not strong enough to produce any sort of reasonable car, let alone a better one.

THE POWER OF NATURAL SELECTION

Fortunately, we're not done yet with our demonstration. There is one more feature I can add in to my program and it turns out this single feature, called Natural Selection, completely changes everything. There's nothing unusual or complicated about Natural Selection. It simply means that any car that's more likely to 'survive' than other cars is quite reasonably going to have more offspring. In our car situation we can think of it in the following way:

Let's suppose the car with the most aerodynamic shape is the fastest and because of this gets to meet and mate with more cars. This is a quite a reasonable scenario and reflects what happens in real life to animals. If you're the first person to get to a limited food source you'll be able to eat all you want whereas later arrivals will have progressively less to eat. And if you're desperate for food and likely to die of starvation yourself, you're unlikely to be in a position to breed new children and ensure they have enough food to survive. Viewed like this, Natural Selection is really quite a reasonable thing to add in and simply reflects the only way things could naturally work if left on their own.

So what happens now?

Wow! What a difference! Within a few hundred generations your Model T starts to look like an E-Type Jag! The combination of Natural Selection and Random Chance working together has the power and the means to produce some awesome cars without any designer hidden anywhere in the system. On their own, Natural Selection and Random Chance are individually as powerless as the elderly woman, but put together they wield truly enormous power.

Again, I want you to be comfortable there are no tricks in here. All I've done is calculate how fast each new car is going to go based on simple laws of Physics — aerodynamics and handling. Then the faster any car goes, the more likely it is that it'll get to the gas station first and top up its tanks, and that in turn means it's more likely to have more offspring. It's as simple as that.

CHALLENGE TO READERS

I actually wrote this computer simulation myself in a mathematical programming language called Matlab and it worked an absolute treat. It was amazing to watch the cars evolve from a Model T shape to an E-Type jag shape before my eyes. Unfortunately, Matlab is a very expensive program and it's not one you can easily share with other people or run on a website, but because it's such a neat demonstration I'd love someone to write it in a programming language that can be shared with everyone. If you can do this, I'd be happy to put a link to your website so you can get all the glory. (My programming is not good enough for me to write it in another language.)

If you try running the simulation a second time you'll find you get a different result for your final car than you did the first time around. It will still be an aerodynamic-looking sports car but maybe this time it looks more like a Ferrari and next time it looks more like a Mustang. This reflects another feature of Random Chance: there's no guarantee where you'll end up or whether you'll ever get the very best design. Random Chance doesn't have any goal in mind and to this extent it's quite legitimate to say it is 'blind'. But despite being blind, Natural Selection slowly but surely improves the product even if there are detours and errors on the way. It's all to do with having enough trials and weeding out the failures by Natural Selection. This means you'll also see some rather unusual-looking cars that were developed and then discarded on the way as newer and better models supersede the old ones.

This is exactly what we observe in the fossil record of animals and humans — a plethora of experiments and trials that last for a while and then are overtaken by newer and better models.

IF NATURAL SELECTION IS SO POWERFUL, WHY DON'T ENGINEERS USE EVOLUTION TO DESIGN THINGS?

About now you might be wondering to yourself: well, if Natural Selection and Random Chance are so powerful, why don't engineers just use Evolution to design things? Good question.

The answer is — they do!

There is a powerful branch of Science which uses genetic algorithms to solve many problems that previously stumped the greatest engineers and scientists. These algorithms work on the exact same principles I used above to evolve our E-Type Jag from a Model T. In fact, my little simulation was just a primitive genetic algorithm.

From wing design to optimizing electronic circuit board layouts to inventing new mathematical formulae, genetic algorithms are a staggering success. If you're interested in seeing some of the uses they've been put to just try Googling 'genetic algorithm' and you'll find there are millions of examples. They're particularly useful when the problem you're confronted with is especially complex with thousands of competing factors mixing together in an unpredictable manner. Exactly the situation animals face in real life.

THE ORDER OF RANDOM CHANCE VERSUS DESIGN

Another thing you'll notice with our simulation is that there's a nice progression from older to newer cars as the simulation unfolds. You never see fully developed E-Type Jags at the 30 per cent stage of the simulation followed by 'half E-Types and half Model T's' appearing at the 70 per cent stage. No, there is *always* a nice progression where bits of previous models are modified and used in future models. This is another hallmark or fingerprint of Random Chance and Natural Selection working hand in hand. Think about it for a moment: an all-wise and all-powerful God could design and create objects in any order he wanted to. He could just as easily have made the E-Type at the same time as the Model T or made the E-Type before the Model T.

In the same way God could have made the jellyfish at the same time as humans or even before us but we never see this in the fossil record. In millions after millions of archaeological explorations around the world we find the exact same progression where bits are taken from old models (animals) and modified and reused in later more advanced models (animals). What's also quite telling is that many of these 'repurposed' components don't function very well in the new models and we often end up with interim designs that don't work too well. They're a slight improvement on the old model but the result is definitely an interim kludge. A real designer would naturally bypass these intermediate steps and go from one model to a better one.

I'm not going to go over the ground so brilliantly covered by people like Dawkins, Coyne and others who have far more knowledge of the fossil record than I'll ever have, so I suggest you get one of their books and see for yourselves just how complete and compelling the fossil record is and how it always follows this process of reusing old components.

THE ORIGIN OF DIFFERENT SPECIES

It turns out we're not quite done with our car simulation yet as there is an interesting twist in the tail. Suppose I tarseal some of the roads in my simulation and make them smooth while I make other roads bumpy, mud-covered trails. What happens then?

Before long I find I end up with two distinct types of car: one low-slung and suited for zipping around on smooth roads and another car with lots of ground clearance and suspension suited for the trails. In short, I get rally cars and race track cars. Before long these cars are so different they can't interbreed. What this means is that we've developed a completely new species! So transmutation is possible after all.

THE TICKING OF THE EVOLUTIONARY CLOCK

You'll recall a key aspect of our car simulation program was that we introduced small random errors in the copying process when we took half the genes from one parent car and half the genes from another parent car. While we can never predict exactly when any particular copying error will occur, because they are random, we can nevertheless observe the overall average rate of these copying errors over long periods of time. When we do this it turns out these random errors — or mutations — occur at a reasonably constant rate in real-life animals and organisms. This provides us with another really strong clue that Random Chance left her fingerprints all over our design.

Here's how it works.

We first compare the 'genes' (or numbers) inside our Model T and our E-Type Jag and count up the total number of differences we notice between them. If we also know the average rate of copying errors between each generation then we can work out the minimum number of generations that must have occurred between our Model T and our E-Type.

For example, if on average we notice 1 copying error every 100 generations and we find there are 20 changes produced by errors between our Model T and our E-Type then we can say there must be approximately 20 × 100 = 2000 generations between our Model T and the E-Type. And if we noticed 40 differences in the 'genes' between two other cars, we'd expect it would take 4000 generations to go from one to the other. Of course this is only a rough approximation because we can only ever talk of averages when it comes to Random Chance. So what do we find when we look at the fossil record of extinct animals?

We find an excellent match between how long ago different animals lived and how many changes there are between their genes. We can schematically show this on the graph below.

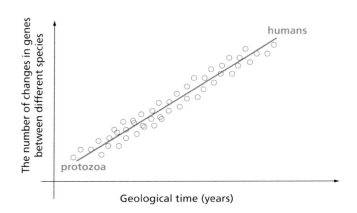

Now this match between the number of genes changed and the time taken doesn't make any sense if animals were created by a designer because a creator could introduce a new species at any time he fancied and in any order he liked. He could also make the next new species as different from the previous ones as he liked. And if that happened we'd never get a nice correlation such as the graph above. Instead we'd end up with a scatter diagram or one with great holes missing in the curve. But we never find this.

What we see is a slow but steady change in a step-by-step fashion with no steps missing. It's as natural and complete as a set of footprints left on a beach. Because the changes we see match the rate produced by the relentless ticking of Random Chance and Natural Selection there's no room left over for anyone or anything else to be involved. In short, Random Chance and Natural Selection not only have the ability to produce all the animals but the method exactly matches the modus operandi of Evolution. It's the *exact opposite* of what you'd expect if a creator was involved.

DIFFERENT TYPES OF HUMAN

Another commonly accepted method of proving guilt or innocence in the courtroom is to rely on DNA evidence. Indeed, DNA evidence is now so powerful and so widely accepted that we have no qualms about sending a person to jail for the rest of their lives if they're linked by DNA to a crime. In many murder or rape cases DNA is the 'gold standard' of proof. Now, if we can accept the reliability and accuracy of DNA in these matters of life and death, surely we can accept what the very same DNA has to say about our own origin. And what it says is truly mind-boggling.

DNA evidence extracted from ancient skeletons proves that we once lived alongside many quite distinct types of humans — Neanderthals and Denisovians to name but two. While previously there was vigorous argument as to whether the Neanderthals really were a separate type of species to us or were instead just deformed versions of modern humans or another type of ape, recent analysis of their DNA shows they were indeed a completely different breed of hominid to us. They're almost like modern humans but not quite. We can be as sure of this extraordinary claim as we can be of any DNA analysis presented in any courtroom. If we doubt this claim then we need to set free everyone convicted on the basis of DNA evidence.

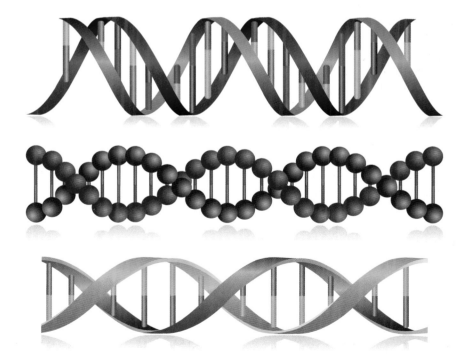

As well as showing we lived side by side with hominids at the same time, this DNA analysis also reveals that we actually interbred with them about 100,000 years ago. Again, we can be as sure of this as anyone can be when they take a paternity test that uses DNA analysis. It turns out that about 4 per cent of the DNA of Europeans comes from the Neanderthal ancestors whereas 6 per cent of Melanesian DNA comes from the Denisovians.

THE SUMMING-UP

Taking all the evidence together, we not only find that Random Chance and Natural Selection left their fingerprints all over our individual design, but additionally our origin follows the exact method and all the stages that are unique to the modus operandi of Random Chance and Natural Selection. If there ever was any interference from an outside 'designer' that was different to Random Chance, we wouldn't get the pattern we observe in the fossil record. Furthermore, despite our natural common-sense inclination, Random Chance combined with Natural Selection proves to be more than capable of pulling off our creation.

We can be as certain we evolved by Random Chance and Natural Selection as we can of our own parents after a DNA paternity test. This conclusion leaves no room for a creator or a personal God. There isn't even any room for a God who might 'direct or control' Evolution because Random Chance and Natural Selection alone not only account for all the facts, any designer would also have to change the fingerprint somehow otherwise he wouldn't have made any contribution.

As the philosopher Daniel Dennett said so eloquently:

> *'Evolution, if properly understood, is completely incompatible with any religion that is predicated on the primacy of humans and on a God who even cares that we exist.'*

Fortunately, this conclusion does *not* mean we can't lead exciting, moral and fulfilling lives. In fact, life without a God pulling the strings turns out to be more exciting than one with a God as we'll see in the upcoming chapters.

CHAPTER 14
WHAT CAUSED IT ALL?

By now you're probably itching to ask me some pretty tough questions. Even though you might accept the arguments about the Big Bang and Evolution, you've still got every right to ask the following questions:

- Even if the Big Bang did occur as you outlined in Chapter 4, what caused the Big Bang in the first place? Surely, we still need a God to make the Big Bang?
- It's all very well showing Evolution works and is sufficient to produce the human race, but don't we still need a God to create the raw materials on which Evolution can work?

This line of argument lit up the internet immediately after Richard Dawkins released his bestselling book *The God Delusion*. Hundreds of thousands of blogs complained Dawkins hadn't disproved the existence of God just because he might have shown humans evolved from simple cells. They accepted his evidence for Evolution but argued you still need a God to create the raw materials on which Evolution worked. According to them, all Dawkins had done was shift the argument back to an earlier unexplained phenomenon that still required the existence of a supernatural intervention — the creation of something out of nothing — and that brought them straight back to God.

The same sort of furore erupted when Stephen Hawking claimed in his book *The Grand Design* that you don't need a God because the laws of Physics can produce 'something out of nothing' all on their own. Again, the internet and newspapers were filled with the obvious counter-argument: 'But who created the laws of Physics? Surely, these wonderfully complex laws that produced the Big Bang are just more evidence of God's design?' In other words, the laws of Physics, like everything else, still need a creator and so we still need a God.

The idea for all these arguments comes from our general everyday observation that 'everything has a cause'. The car crashed because it was going too fast. The balloon popped because Johnnie pricked it with a pin. Suzie won the Olympics event because she was naturally gifted and trained harder than everyone else. And so on. It just feels right for 'everything to have a cause'. And if 'everything has a cause', by extension the Big Bang and the Universe must also have a cause.

But is this true? Does everything have a cause?

We begin our answer to this question with the obvious observation that we only have two possible choices. Either:

1. Everything has a cause

   or

2. Some things don't have a cause.

Let's start with the first option and see where it gets us.

OPTION 1: EVERYTHING HAS A CAUSE

If everything has a deeper cause then we must get an infinite regression of never-ending causes. Here's how it works. Suppose we observe something called A, then by our rule it must have a cause which we can call B. But as soon as we find B we must ask ourselves, 'What caused B?' According to our rule B must have a deeper or prior cause which we then call C. Once again we need to ask ourselves, 'What caused C?' It's quickly obvious the claim that 'everything has a cause' immediately gets us into an infinite chain of things being caused by other things without any possibility of an end in sight.

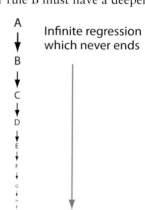

This state of affairs is a bit like the ancient myth which claims the Earth is supported on the back of a giant turtle. Unfortunately, that doesn't really solve anything because we then need to ask the obvious question, 'What's holding up the turtle?' You need another turtle underneath it. Before long you need an infinite number of turtles holding up other turtles in a stack that goes on forever. The other consequence of accepting 'everything has a cause' is that God himself needs a cause and this means he can't ultimately be responsible for you or the Universe. After all, if everything needs a cause then God also needs a creator or a cause.

The only way around this problem is to modify our first claim and say, 'Everything needs a sufficient cause.' In this case we could claim God is a special case because he is so powerful he can just exist all on his own. Under this definition, God is defined as being self-sufficient. But all that statement really means is we're now saying, 'Some things such as God don't need a cause.' They just exist. If we accept this idea we now have a very long chain of 'turtles', but instead of going on forever, there's a box at the bottom on which everything rests. But the existence of this box immediately forces us to reject our first option because the only way it can be at the bottom all by itself is if 'Some things don't have a cause' after all.

Something without any cause — it just is.

OPTION 2: SOME THINGS DON'T HAVE A CAUSE

We've just seen the only way we can have God at the bottom and being the original cause of the Universe is for there to be at least one thing which doesn't have a cause. There's no other way around it. But if there's one thing in our 'box without a cause', what's to say there aren't two or 50? And how do we know if it's the God of the Bible or the Qur'an or Zeus that's lurking inside this box?

Here's where things get rather interesting. It turns out there are many items in our 'box of without causes'.

Consider a rigid lever which pivots about a fixed point. If the right side of the lever is twice as long as the left side then the right end will always move twice as far as the left end. There is nothing anyone can do about this. That relationship between the length of each arm and how far it moves 'just is'.

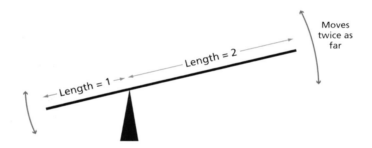

It doesn't need a designer or a creator to make this happen. In fact every other possibility, such as the right side moving five times as far, just leads to an impossible situation. Asking that to happen would be like asking me to be both taller and shorter than myself at the same time.

Our observation about the length of levers and how far they move can be written out mathematically in the following way, which will hold for all levers regardless of how long each side is: $D2 = D1 \times L2/L1$

What's important about this law is that it *must* be true all by itself. In the same way many other laws exist without the need for any creator; they're necessarily true all by themselves. You'd probably never guess Pythagoras' rule $c^2 = a^2 + b^2$

just by looking at a right-angled triangle, but it's *also* necessarily true.[1] It would be true if there were no triangles in the Universe or even if there wasn't a universe at all! Not even God could change Pythagoras because if he did we'd end up with situations that were impossible, not just physically impossible but logically impossible — like me being both taller and shorter than myself. Even if we think God could break the laws of Physics in an event we'd call a miracle, it's still impossible for him to break the laws of Logic. If you're not convinced about this then ask yourself the following question: 'Can God create a problem he can't solve?' Whatever answer you come up with, yes or no, you're forced to admit there is something God can't do.

There are two surprising things about these necessarily true laws:

1. There are a *lot* of them.
2. They have unexpected consequences.

Each of these necessarily true laws give birth to thousands of other less obvious and more complex daughter rules, all describing things that are equally necessarily true in any universe. For example, we can quickly derive a 'daughter' rule describing the forces at the end of levers from our first rule about how far levers must move. And that new rule concerning forces in turn allows us to work out how gears work in the real world. Before long we find there is a massive network of interconnected rules which are all necessarily true and which give our Universe certain features it *must* have. Features that no supernatural agent could change and which don't rely on a God for their existence.

One of the most bizarre predictions made by Mathematics is that a completely empty vacuum will on occasion spontaneously produce an entire universe out of absolute nothingness in a process described by quantum mechanics. This is such an important discovery I've devoted the next chapter to it.

---

1. Strictly speaking Pythagoras' theorem only holds true for a triangle on a flat plane, but this doesn't matter because we can derive other formulae for any other shaped surface on which triangles are drawn.

CHAPTER 15
QUANTUM MECHANICS

'Anyone who is not shocked by quantum mechanics cannot possibly have understood it.' — Niels Bohr, winner of the Nobel Prize in 1922 for contributions to quantum mechanics

EINSTEIN'S ARCH ENEMY

Of all the people who've ever lived, you'd think Einstein would be the *one* person who'd be able to cope with finding out the Universe operates in some bizarre manner. After all, he was the person who discovered solid lumps of matter (m) like rocks and stones were nothing more than hot shapeless Energy (E) that had crystallized into existence out of nothing according to his famous formula $E = mc^2$. He was also the same person who said the flow of time wasn't fixed but actually slowed down if you moved very fast or experienced a strong pull of gravity. It didn't bother Einstein that a single second for one person travelling close to the speed of light would literally take thousands of years for another group of people watching on.

Such was the power of Einstein's mind he could take these counter-intuitive ideas in his stride even though the rest of the scientific community needed another 30 years before they caught up with him. (If you think these ideas still sound unlikely, we know they're true because the atomic bombs that exploded over Nagasaki and Hiroshima were designed using Einstein's equations. The devastation that followed was surely the most powerful proof of any mathematical formula in humankind's history.)

Now if Einstein could accept all those weird things without even batting an eyelid, you'd expect it would take something *really* strange to stop him in his tracks. Quantum mechanics was just such a thing. It is so shocking Einstein devoted the last 40 years of his life trying to find fault with it, despite the fact he was the very person who helped discover it in the first place. Another 60 years later and we find quantum mechanics has still passed every single test we can throw at it with flying colours. Some even claim it is now the most tested and proven description of Reality we have. All our modern electronic equipment like DVD players, computer chips, CCD cameras and TVs were invented because engineers took the bizarre predictions of quantum mechanics at face value and built equipment according to its laws. To their utter amazement these strange electronic marvels actually worked.

So, without realizing it, we all rely on quantum mechanics in our everyday lives. Despite this success, what quantum mechanics says about Reality will shock you to your core, or at least it should if my explanation gives you even a hint of what it says about the fabric of Reality.

THE TEASPOON AND THE MOUNTAINS

'I think I can safely say no one understands quantum mechanics.'
— Richard Feynman, 1965 Nobel Prize for contributions to quantum mechanics

If academic giants like Bohr, Feynman and Einstein who specialized in quantum mechanics say no one can understand it then I've got a pretty big task on my hands trying to explain it in a simple book like this. Worse still, I'm going to try to do so in just a few pages and without using any of the Mathematics it's normally written in. I'll make my task a little bit easier by only attempting to explain 'what' quantum mechanics says about Reality rather trying to explain 'why' quantum mechanics has to be true.

Let's start by imagining there's a huge, heavy rock sitting firmly on the ground in front of us. In the distance many kilometres away from us there's a man with a teaspoon and a coin. He's so far away from us we can't even see him. This fictional man is engaged in a rather strange task; he tosses a coin and if he gets a 'head' he adds exactly one teaspoon of soil to a pile of dirt on his right and if he throws a 'tail' he adds exactly one teaspoon of soil to a pile of dirt on his left. Because he's an imaginary man we can make him work as quickly as we like, and so to speed things up I'm going to make him toss his coin 100 billion times every second.

At this blistering rate the two piles of soil grow quite quickly even though he's only using a teaspoon. Before long we begin to see these piles from where we are standing even though they're so far away. First they become small hills and then with enough time they become decent-sized mountains. Because the coin he uses is perfectly balanced we notice the two piles of soil seem to grow at exactly the same rate. This is not surprising, because after many trillions of throws with a perfectly fair coin we'd expect very close to half the throws to be heads and half the throws to be tails. Any extra head or tail, an extra teaspoon here or there out of the trillions and trillions of teaspoons needed to make a large mountain isn't going to be noticeable kilometres away.

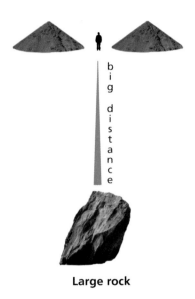

Large rock

While all this tossing of coins and adding soil with teaspoons is going on in the distance, our task is to measure the exact speed and position of the large rock that's sitting in front of us. To make our lives easier there is absolutely no force acting on the rock. The ground is completely still and there's not even a breath of air to disturb it. The rock is completely and utterly isolated from any external influence. Now this is where things get a little strange. Let's suppose the exact position of our rock each time we measure it is somehow influenced by the relative sizes of the mountains being built by the man in the distance with the teaspoon according to the following rule:

Rule 1: The position of the rock depends on the relative sizes of the mountains

1. If the mountain on the left is bigger than the mountain on the right then we find our rock has moved to the left. The bigger the left-hand mountain is the further our rock has moved to the left.
2. If the mountain on the right is bigger than the left mountain then we find our rock has moved to the right. The bigger the right-hand mountain is the further our rock has moved to the right.
3. If the two mountains are the same size then the rock stays perfectly still.

Despite this rather strange rule, we find our rock hasn't moved each time we measure it because from our distant vantage point, the two mountains always appear exactly the same size. Again, that's not surprising considering how far away we are and how many teaspoons are required to make up each mountain. So despite making the behaviour of everyday objects obey Rule 1 we find our normal experience of measuring things hasn't changed. But let's see what happens when we add in one last rule which when combined with our first rule describes the essence of quantum mechanics.

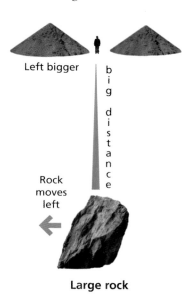

Left bigger

b i g

d i s t a n c e

Rock moves left

Large rock

Rule 2: The distance you stand from the two mountains depends on how small the object you are trying to measure is

For big objects like rocks or people we stand hundreds of kilometres away from our man with his teaspoon and coin, but as the objects we want to measure get smaller and smaller we are required to move closer and closer to him. For very small objects the size of atoms we must stand a few metres away from him and watch each time he tosses the coin rather than waiting till he's tossed trillions of coins. Because we're so close we can now see the piles of soil while they are still very small and at this scale we notice they aren't always the same size. Sometimes he tosses three heads in a row, and therefore according to our first rule, our atom is three 'spaces' to the right the next time we measure its position. At other times we'll find he tosses a few extra tails and so we find our atom has moved to the left. In this way we find our atom jiggles backwards and forwards about its original position each time we measure it.

Of course these jiggles back and forth leave the atom in roughly the same place overall because on average he still throws as many heads as tails. But the point is, at this small distance we can definitely see two piles aren't always the same size, as first one then the other is bigger. On very rare occasions our man might even throw 10 heads in a row and so in that case we're surprised to find our atom has moved quite a long way from where we were expecting it to be. Once again, I must stress that the position of our object, whether a rock or an atom, depends only on the random toss of the coin and has nothing to do with any forces acting on it.

With these two rules in place we find normal-sized objects like people or even viruses behave as we expect them to. Items stay where they're put and they have a definite speed or position each time we measure them. But for very small items like atoms or electrons (in the quantum realm) we find they don't have a fixed position or speed. At one instant they're in one place and at another instant we find they're somewhere else. This has nothing to do with our measuring ability and it's not caused by any outside influence. At the very small scale the world becomes a weird place where nothing is certain. At this tiny level an object doesn't even have a definite position until you measure it. All you can do is work out the probabilities it will be in certain places according to your coin tosses. Most of the time it will be close to where you left it but sometimes it will have jumped far away.

The only reason we see things having fixed positions as we normally do in our everyday world is because at this large scale all the tiny jiggles and movements have averaged out and our view of the 'mountains' is poor.

It turns out our two rather strange rules capture the essence of quantum mechanics and have profound consequences, including the prediction the Universe can start all by itself out of absolute nothingness. But before we talk about this shocking prediction about what is in the box at the bottom of our turtles, I want to show you how quantum mechanics works in your daily life as this will help make clear what is so strange about this branch of Physics.

SOME PRACTICAL APPLICATIONS OF QUANTUM MECHANICS

As crazy as our two rules sound, their effects are felt in our daily lives because they make modern computer chips possible. In the diagram on the right we see a semiconductor with lots of electrons visible on the left-hand side. The right-hand side of the semiconductor is separated from the left by a barrier that is truly impenetrable for electrons. In normal everyday life this means the electrons will always be contained in the left-hand side. But according to our two rules, the position of tiny objects like electrons is never fixed and they jump around quite a bit because we're so close up to our man with his teaspoon. Sometimes the electrons turn up to the left of where we expect them to be and sometimes they turn up to the right of where we expect them. On rare occasions there is a small probability they will have jumped a substantial distance, which is equivalent to throwing 20 heads in a row.

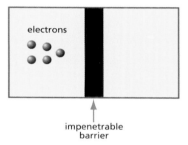

Semiconductor

electrons

impenetrable
barrier

When this happens we find some electrons unexpectedly turn up in the right-hand side of the box even though the barrier between the left- and right-hand side of the semiconductor is totally impenetrable. To get from one side to the other the electrons don't travel through the barrier in the normal sense we're accustomed to in everyday life, but instead they instantly jump from one side to the other without going 'through' the barrier in the middle. At one moment they're in the left-hand

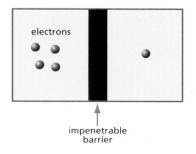

Semiconductor

electrons

impenetrable
barrier

side and the next they're in the right-hand side without ever going through the middle. We *never* catch them halfway through their journey in the middle of the barrier. In other words, if we throw 20 heads in a row and our electron needs to jump 20 places to the right, it doesn't do this by moving first one place, then two places and so on up to 20. It simply disappears from the Universe at the start position and pops out again at the new position 20 spaces away without ever being in any place in the middle.

Of course this sounds impossible and there is nothing in our normal lives that can prepare us for items moving in a discontinuous manner. To our ant-like brain it sounds impossible but it's what actually happens in the physical world and it's all to do with our 'little man tossing the coins'. If it didn't happen, semiconductors wouldn't work and neither would all your electronic gadgets, because they rely on electrons 'jumping' from one place to another according to our coin toss rule.

This example illustrates one of the weird aspects of quantum mechanics: the position of tiny particles like electrons is not smooth — they do not move from one place to another in continuous fashion in the way we observe large objects such as bacteria or viruses. Instead their position is completely undetermined until we measure it the next time and in between our measurements no one can say how it got from one place to the next. This indeterminacy is not due to our lack of knowledge or our lack of equipment to measure them — they simply don't have a position between throws. It would be similar to asking whether a coin was 'heads' or 'tails' between throws. That's a question which doesn't make any sense.

Einstein and many other scientists hated this discovery and tried as hard as they could to get rid of what they called 'This damned quantum jumping', but the more they looked the more they found it permeated every aspect of Reality. If you want to explore this jumping phenomenon inside electronics in more detail you can Google 'quantum tunnelling', which is what scientists call it.

The same 'jumping' also explains why certain elements such as uranium are radioactive. Sometimes the inner constituents of these large atoms randomly find themselves outside their normal positions and as a result they're no longer held together by the inner nuclear forces and so they get violently ejected, producing radiation. One of the earliest triumphs of quantum mechanics was that it accurately predicted how often this ejection would happen for each element, which in turn predicted their half-life and the amount of radiation they emit.

THE SCARY PART THAT EINSTEIN HATED

Our two rules describe how our Universe actually works. This means at the very bottom of our long line of 'turtles' is a final box which has nothing inside it but pure Random Chance. We can think of this box as containing a coin being tossed thousands of times to determine the state of each small component of the Universe. In fact there's more than one coin for each particle in the Universe because particles have many properties such as their position, speed, momentum, polarization and so on. Each aspect for every particle requires a separate coin.

Of course there aren't actually 'coins' spinning in there or anyone throwing the coins — just random probabilities coming out all by themselves which determine what is going to happen to each microscopic component of the Universe. No one is making these probabilities — they are just necessarily there like our rule about how far levers move. The key point is, they are absolutely, completely and utterly random. It's only because there are so many of them spinning around that they average out to give us our normal everyday experiences which look so predictable. At our large-scale view standing far away from the two mountains, everything seems stable and life works according to the normal rules of Physics. But these rules are themselves the result of millions of statistical probabilities added up like so many teaspoons.

Einstein hated this discovery, which is why he famously said: 'God does not play dice with the Universe.' (Whenever Einstein used the word 'God' he always meant 'the rules of the Universe' as he didn't believe in the existence of a personal God in the normal sense. Niels Bohr replied to Einstein's quote by saying: 'Do not presume to tell God what to do.'

PARTICLES FROM NOTHING

This finally brings us to the nub of what quantum mechanics tells us about Reality. The most significant consequence of these 'little coins' spinning in the bottom box is that it explains where the Universe came from — where you came from. I mentioned before the coins don't just determine the position of each atomic particle, they also determine how much energy everything has. If we apply these rules to a completely empty vacuum containing absolutely nothing, we find sometimes the vacuum has quite considerable energy — equivalent to tossing lots of heads in a row. This surplus of energy can only exist for an extremely short period of time before other 'coin tosses' balance things out and we get back to zero energy. But during this short time Einstein's equation $E = mc^2$ predicts this temporary surplus of Energy (E) will turn into a solid lump of matter (m). When that happens a particle and its anti-particle suddenly 'pop' into existence out of nothing. (An anti-particle is just the mirror image of its partner particle and they're always produced in pairs like this as this ensures their electric charges balance out. For example, the anti-particle of the negative electron is called a positron.) Usually, these particle pairs disappear again almost instantly because the next coin tosses balance things out again, but they do genuinely exist for a fraction of a second. This means a completely empty vacuum is really a seething mass of particles popping in and out of existence at a furious rate. It's only because they are so small and because they exist for such a tiny fraction of a nanosecond that we aren't aware of them. But if we look extremely closely on a very tiny scale we can actually catch them in the act of appearing and disappearing. (If you want to find out more about this then you can Google 'virtual particles' and the 'Casimir effect'.)

A good way to think of this process is to imagine that the 'empty vacuum of nothingness' is like a thin sheet of rubber making up a balloon. Each time we toss a coin we stretch a tiny piece of the 'rubber', and while it's stretched and under tension, a particle and its anti-particle are briefly visible. But as soon as the next coin toss balances things out, the 'vacuum rubber' relaxes back into place and the particles disappear. This constant popping in and out of particles provides the pressure which keeps the balloon inflated. (Scientists call this pressure the vacuum energy.)

Now here's where things get incredibly interesting!

A UNIVERSE FROM NOTHING

Occasionally, we'll toss many heads in a row, which causes the rubber to become so stretched a small tear forms. Now the rubber can no longer snap back into place but instead a rip rapidly expands through the fabric with such violence the balloon pops. It's a one-way process that can no longer be reversed. As soon as this happens the hidden energy of the vacuum is released, producing first space and time, and then the huge explosion we now know is the Big Bang. Initially, this energy drives the explosion apart before the temperature cools down as space expands, which then allows particles of matter to crystallize out from the energy according to Einstein's equation $E = mc^2$. This apparently ultimate 'free lunch' is a necessary consequence of all that 'coin throwing'. It just has to occur whether we like it or not.

Stephen Hawking and others have shown it's more likely for a universe to come out of an empty vacuum like this than it is for an empty vacuum to always snap back into a smooth sheet of rubber and thus stay empty forever.

Another consequence of all this coin tossing is that our entire Universe is likely to be just one of many other bubbles that burst. We can now imagine the vacuum producing lots of little bubbles which spontaneously expand to form their own complete universes separated from ours. Even though our Universe is trillions of miles in size with trillions of stars, it is but one universe in an endless mosaic

of other universes, all completely separated from each other. We call this massive mosaic the Multiverse. In the image on the right, each bubble represents an entire universe complete with galaxies and stars. Some of these universes will be similar to ours and others will be completely bizarre. Suddenly, our place in the Multiverse has become even smaller.

THE FINGERPRINTS OF CHANCE

In the same way we saw how Random Chance left her fingerprints all over Evolution in Chapter 13, so too all this coin tossing must leave very specific fingerprints on our Universe today if quantum mechanics did indeed produce our Universe from a rip in the vacuum.

The first fingerprint we must see is that all the energy of our entire Universe must add up to *exactly zero* when we use the correct accounting system. In cosmology we need to count all the energy of gravity as being negative and all the energy due to the mass of stars, the heat of the stars and the explosion as being positive. This requires enormous sums and it would be an impossible fluke if all these huge numbers balanced out to exactly zero. But that's precisely what we find when everything is added up. (Scientists summarize this exact balancing out by saying our Universe is 'flat'.)

The second fingerprint that ought to be left over if the Universe popped out of a 'vacuum of nothingness' is that there should be tiny bumps in the heat of the Big Bang we see left over today. These tiny bumps in temperature must have exactly the right variation in temperature and be of exactly the right size as they are spread out through the entire Universe. It's like specifying how many red, black, green and blue Smarties there are in a packet, how big each Smartie should be and where we should find them. That's a pretty tough prediction to make and here's why it happens.

If you remember back in Chapter 4 we saw the background radiation spread throughout our entire Universe is 2.725 degrees Kelvin (−270.425 degrees Celsius) which precisely matches the prediction of the heat left over from the Big Bang (remember the coffee cup cooling down?). But if our Universe was indeed caused by 'coins' being tossed and ripping the vacuum, there should be tiny variations in this temperature depending on the pattern of coins thrown. Sometimes there'll be two 'heads' in a row, sometimes five heads in a row while at other times we'll throw more tails. How often we get each different pattern of heads and tails determines the size and colour of the Smarties or, in other words, the size and distribution of tiny variations in temperature of the background radiation.

We expect these variations to be around 0.0002 degrees and to be spread in a distinctive pattern.

Not surprisingly, there was frenzied excitement in the scientific community when a recent NASA space probe (called WMAP) showed temperature variations across the entire Universe which exactly matched all the predicted features of the fingerprints of our coins. It's as certain as any DNA match in a court of law.

The implications of this image are so extraordinary and so profound I've even printed it out and hung it on my living room wall. For the first time in humankind's history we can look directly into the bottom box sitting underneath all those turtles  and see exactly what's in there. There we see the imprints of **individual coin tosses from the very start of time!** If you get the significance of this achievement it ought to give you goose bumps. We are left with the beautiful ... the simple ... 'magic' of the spinning coins. There is no other word for it. Don't forget there isn't anything causing these random numbers and I've only used the analogy of spinning coins to help explain the utter randomness coming out of the bottom box. And even if you say to yourself, 'Well, God is tossing the coins', this adds absolutely nothing. In that case God is totally and utterly without influence because each throw is completely unaffected by any previous throw or who or what is doing the throwing. God would have no more effect on the outcome than a referee does when he tosses the coin before a football match to decide who kicks off first.

In closing, don't worry if you've found this chapter strange or difficult. After all, the greatest mind in the world, Einstein, spent 30 years trying to grapple with it. With a bit of luck you might be motivated to read some popular introductory books which explore this fascinating topic in more detail. In those books you'll find even more magical properties like 'entangled objects' which can be separated by millions of kilometres while still being instantly aware of what their partners are up to. You'll discover that Reality doesn't make its mind up until you look at what it's doing. All this magic is firmly founded in Logic and produced by rules which are necessarily true. I wish you well on that journey.

CHAPTER 16
IS THERE LIFE AFTER DEATH?

'Millions long for immortality who don't know what to do with themselves on a rainy Sunday afternoon.' — Susan Ertz

Is there life after death? This is probably the most important question we can ever ask ourselves because it determines our place in the cosmos and our status as humans more than any other question. If we simply disappear and vanish into empty nothingness when we die then we are entirely different beings than if we go to Heaven or are reincarnated. Nothing affects our status more than whether we have a soul that continues on or doesn't. This question also distinguishes Religion from Philosophy because all religions require some sort of soul or spirit to continue on after death whereas Philosophy doesn't. But how can we possibly hope to answer this question without first dying and then coming back from the dead? One solution would be to find a way to look beyond the veil of death without ever having first crossed over there.

Fortunately, there is a way we can do just that!

THE MOTORCYCLE ANALOGY

I think the best way to start is by using a non-human analogy so we can take some of the emotion out of this deeply personal question. In this case I'll use an analogy based around my motorcycle. Despite the rather flippant, tongue-in-cheek tone I use, it will serve us well as it illustrates the key points.

I love motorbikes and at last count I've owned 29 of them, including my all-time favourite, the Suzuki Hayabusa. Over the years there's been one very puzzling thing I've noticed with every one of them. Whenever I turn the engine on I sense a mysterious thing I've decided to call the 'sound' of the bike, and whenever I turn the engine off this mysterious phenomenon called 'sound' stops.

Now this is all rather strange because my bike is made out of carbon fibre and exotic alloys and yet the sound I hear clearly isn't made out of any of these things. I can touch and feel the bike's physical presence with my hands but I experience sound quite differently. I can even hear it when I'm in another room and my bike is in the garage. It's almost like telepathy. I can tell when my bike is alive or 'dead' simply by listening for this sound thing.

In my naivety I conclude the bike must be made out of two totally distinct things:

1. A 'physical presence' — its nuts and bolts.
2. A separate independent and mysterious thing I've called its 'sound essence'.

My belief in this separate sound essence is further confirmed because even if I pull my bike apart and inspect all its internals in microscopic detail — the pistons, valves and crankshafts — none of these physical things seems to make the mysterious sound I hear when the engine is running. I can tap or bang each component and no matter what I do with them, none of them makes the sound I hear when the bike runs. Even if I add up all the individual sounds that come from each of the individual components I still don't get anything vaguely resembling the sort of sound I get when the engine is alive.

The only conclusion I can come to is there must be some sort of mystical, mysterious essence that accompanies my bike when it's alive. Indeed, every bike I've owned seems to have its own unique sound or 'personality'. Some are temperamental and make rather unhappy sounds until the engine is revving beyond 10,000 rpm, while others are quite content to burble along slowly. It seems the bike's sound is what separates a bike that's alive from one which is not working or 'dead'.

But one thing still worries me: this connection between the bike's physical presence and its sound is so inextricably linked I just can't be sure the sound isn't actually produced by those nuts and bolts after all. The sound and the engine being alive *always* start and stop at exactly the same time. Maybe the sound I hear is produced by its inner metal components despite all the indications to the contrary.

This possibility raises an important point: the more intimately connected these two things are — the running of the engine and the sound — the less inclined I am to believe they are separate unrelated things. If adjustments to the bike (for example, its rpm) produce changes to the sound that *always* happen in a 100 per cent predictable way, then I'm quite within my rights to suspect they're related to each other after all.

BRAINS VERSUS SOULS

Of course I don't really think about my bike like this and I'm certain its sound is produced entirely by the engine running. The reason for belabouring this analogy is that it has many nice parallels with our brains and our souls. But before we go any further we need to define what we mean by the word 'soul', otherwise we'll get ourselves hopelessly confused.

For the sake of this discussion let's say our souls are the same as our personalities and our subjective experiences (we can always refine our definition later if that's helpful). Using this definition our souls contain all our thoughts, feelings, moods and the very 'essence' of who we are. When someone talks about my personality they also talk about how I behave, how I react and what sort of person I am. Am I kind and loving? Fun and crazy? Spontaneous? Honest? Driven? Relaxed? We also need to include all my own internal feelings and sensations such as how I feel when I fall in love, experience fear, taste chocolate, hear music, or feel good about myself. We can wrap all my thoughts, my consciousness, my feelings, my passions and who I am into one big bundle and call it my soul. It not only includes what I feel but also makes me the unique person I am. It's what separates me out from every other person. Given this definition, what is our next step?

THE ANGRY DENTIST AND THE GAMBLING PRIEST

We can take our cue from the motorcycle analogy. What raised our suspicions that the bike's engine was producing the sound was to observe a direct and unequivocal connection between the engine's revs and the sound we heard. If we turned the bike off the sound stopped. If we revved the engine up the sound got louder and higher-pitched. If we ever noticed the same type of correlation with our brains and our personality, we should be equally suspicious they are also linked. And this is exactly what we observe.

Consider the following true story about a dentist who had been happily married to his wife for 23 years. By all accounts he was a devoted father and husband: kind, caring and a pillar of the community. But all of a sudden his personality began to change. Whereas before he was a patient and tolerant man, now he began to lose his temper quite frequently. Even small things would cause him to fly into an uncontrollable rage. Within a few month things deteriorated to such an extent he began to get physically violent with his wife and work colleagues. Eventually, the police were called and he was charged with a serious count of domestic violence.

Fortunately, his local doctor suspected something was amiss and ordered a brain scan. This revealed a tiny tumour growing in a part of his brain which linked the rational frontal lobes of his brain to the primitive emotional circuits. The tumour was interrupting the normal link between these two modules, which 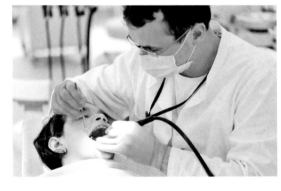 meant he wasn't able to control his instinctive primitive reactions. As a result his entire personality, his soul, changed. It wasn't because the cancerous tumour was annoying him or making him angry, but rather because the wiring between his two modules was preventing one part of his brain from exercising restraint over the other part.

As soon as the cancer was removed the dentist went back to being the same loving, kind and tolerant man he always was. There was the same direct link between his brain and his personality as there was with my bike's engine and its sound.

This is not an isolated case. We notice this same inextricable link time and time again. Let me give you one more example as it has implications for our own moral responsibility and the issue of Heaven and Hell. Consider the real-life case of the British priest who according to all who knew him had led a blameless life of piety and integrity. Unfortunately, he suffered a heart ailment and so his doctor prescribed a particular heart drug. Within weeks the priest had an overwhelming urge to gamble. So bad was this urge that he secretly stole from his church and gambled away his entire life's savings and those of the church before it was discovered what was causing the problem. A simple chemical had temporarily changed how the wiring in his brain functioned, and with that his entire

moral and ethical essence. As soon as he was taken off the drug he returned to his normal pious behaviour — although the issue of all the missing money remained.

This raises the question: if the priest had a soul which was separate from his brain, how on Earth did the drug affect the very essence of his soul and his personality? Everything the priest stood for and believed in was completely changed by the drug. If his personality and his soul, his ethics and morals were separate and independent from his brain, why were they so totally changed by it?

These are but two simple examples of similar observations neurosurgeons and neurologists notice every single day. A bump on the head or a stroke can produce remarkable and permanent changes to our souls, with hardly a single aspect of our personalities that can't be affected by changes to our brain. For some fascinating insights into how the brain affects our personalities you'd be hard pressed to do better than read the bestselling book *The Man who Mistook his Wife for a Hat* by Oliver Sacks. This beautifully written book illustrates with interesting clinical cases what happens to our personalities when various brain modules are unfortunately disrupted by disease or accident.

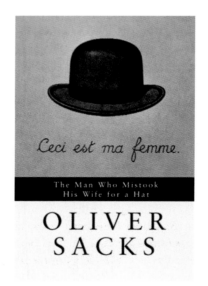

Ceci est ma femme.

The Man Who Mistook
His Wife for a Hat

OLIVER
SACKS

NO CENTRAL COMPUTER

The reason why changes to our brain produce these changes to our personality is now relatively well understood. Our brain is constructed out of many different 'modules' which all work together and combine to produce our personality. There's no central 'CPU' inside our brains like there is inside a computer. There is no place in your brain where 'you' are. Instead, there are hundreds of different modules inside your brain each looking after distinct aspects of your personality. How these different modules combine and interact with each other is what determines your personality. That's why alcohol or drugs affect your behaviour so much. These chemicals affect different modules in different ways. Alcohol, for example, turns off the 'logical restraint' modules while stimulating the emotional ones and so you behave differently. To some extent you temporarily become a different person under the influence of alcohol.

ANAESTHESIA

This observation that the brain affects our personality has its ultimate expression when a patient is anaesthetized and their brain is turned down to a slow state. Even though anaesthetized patients' brains are still functioning and quite alive, the mere presence of a few chemicals eliminates their personality completely. There's 'no one' home. There is nothing left over.

If you've ever had a general anaesthetic you'll know there's just an 'empty nothingness' between when you went under and when you woke up. But this raises the obvious question:

If you had a soul or a spirit, where was it and what was it doing when you were anaesthetized?

If we truly believe in a soul that is separate and survives our physical body, a belief called dualism, then there must be something left over that's completely and utterly unaffected by this general anaesthetic. But I've yet to find anyone who is able to define exactly what this supposed 'leftover' is.

'Since our inner experiences consist of reproductions and combinations of sensory impressions, the concept of a soul without a body seems to me to be empty and devoid of meaning.' — Albert Einstein

NEAR-DEATH EXPERIENCES

By now you might be objecting that this business with anaesthetics is all very well and good but what about so-called 'near-death experiences' where people see tunnels leading into bright lights and so on? Once again, these experiences turn out to be little more than by-products of our brain.

Under extreme situations where the brain is fighting for its life it tends to go through a well-defined process of 'shutting down' to conserve its precious neurons. Fighter pilots under high g-forces develop tunnel vision as the brain conserves precious resources by shutting down the periphery. Boxers in a ring experience all sorts of hallucinations and visions that often seem more real than real to them as their injured brain struggles for survival. Muhammad Ali talked about seeing bats sitting on the ropes and playing trumpets when he was hit. Images so vivid and convincing they are indistinguishable from Reality itself — like a dream that seems as if it really happened to you long after you've woken up.

Researchers have even tested these near-death experiences by placing a large message suspended upside down from the ceiling right over the operating table so that anyone looking down from above could easily read the hidden message. Despite reports of near-death experiences in these rooms where patients report looking down and seeing themselves lying on the operating table no one has yet seen these hidden messages.

WHAT ABOUT AN 'ESSENCE' OR A 'SPIRIT'?

But maybe you object that my definition of a 'soul' is too narrow. Maybe you say there is still something different from our 'personality or soul' which survives both an anaesthetic and death. Maybe it is something altogether different called our 'essence' or 'spirit'. The problem with the words 'spirit' or 'essence' is that they carry a lot of historical baggage that can unconsciously trick us into thinking there's something there even when there isn't. So to be on the safe side, let's make up a neutral word that has no previous historical baggage and call whatever we think is left over when we're dead or deeply anaesthetized our 'ziffle'.

The questions we must then ask ourselves are:

- What is this ziffle and what does it do?
- What evidence do we have to say it exists?

Unless we can define what this ziffle is we are making Neither statements which are no more meaningful than other Neither statements like: 'His weight was a nice green colour.'

While that sounds like a sentence and has all the structure of a sentence, it actually means nothing at all. So what exactly is it? This 'essence' can't have anything to do with our personality because we've already shown that is determined purely and totally by our physical brain. And we can't just say it's 'everything which is left over when we die' because then we've simply uttered a tautology. It would be like saying 'pixies are all those little men living in my garden I can't see'. That definition doesn't make pixies real or prove they exist.

I welcome any reader to post on my website blog (www.taatf.com) any definition that conveys unambiguously what they think is this ziffle.

Next we turn to the question of what evidence we have for the existence of this ziffle. We can believe in electrons even though we can't see them because we have thousands of other independent facts like the existence of electricity, TVs and so on which support a belief in their existence, but what is the evidence for the ziffle? By now we have accumulated enough examples of genuine errors in the traditional religious texts to know we can't rely on them to be the sole reason we believe in the existence of the ziffle. So what other solid evidence do we have in the Green region of our Compass of Truth that makes us believe in the existence of a ziffle? If we don't have any evidence then we have to accept it is just a fiction like Santa Claus.

Again, I welcome anyone to post their evidence for the existence of a ziffle on my website (www.taatf.com)

It seems the only reason people *really* believe in a ziffle that survives death is because they believe it gives life more meaning and value. Because then 'bad people' would get punished and good people would be rewarded. Because we'd be able to meet again with our loved ones. While those might be nice ideas, this doesn't mean that's how the Universe actually operates. Fortunately, we'll see in Chapter 18 that life can have even more meaning and value even if it ends with death rather than carries on forever.

'If you think life is fair then you've been seriously misinformed.' — Eleanor Roosevelt

BRAINS VERSUS SOULS SO FAR

- At first glance our brains and our souls seem vastly different things and yet they also appear to be inextricably linked.
- Changes to our brain produce changes to our personality. Just one example: a priest gambled away church funds and his life savings because a simple chemical temporarily rewired his brain functioning. If the priest had a soul separate from his brain, how could the drug have affected it?
- If we equate a soul with personality, then our brain produces our soul. This means there is no soul that survives death.
- We lack any definitive definition as to what else is 'left over' (ziffle) that is unaffected by a general anaesthetic or death.
- We lack any independent evidence for the existence of a ziffle.
- It seems the main reason for believing in 'life after death' is because people think it will make their lives more meaningful.
- We'll see in later chapters how life can have even more meaning if it ends at death rather than carrying on forever.

Do not pass by my epitaph, traveller.
But having stopped, listen and learn, then go your way.
There is no boat in Hades, no ferryman Charon,
No caretaker Aiakos, no dog Cerberus.
All we who are dead below
Have become bones and ashes, but nothing else.
I have spoken to you honestly, go on, traveller,
Lest even while dead I seem loquacious to you
— Ancient Roman tombstone

CHAPTER 17
IT'S ALL TOO BRIEF

'Infinity is a very long time, especially towards the end.' — *Woody Allen*

One afternoon my daughter Nicky came over for lunch with her two young children Zoe and Thomas. We spent a lovely time chasing each other in the long grass, excitedly looking for birds' nests in the trees and playing on the swings in my bottom paddock. At the end of the afternoon as Nicky lay on a large sofa to grab a moment's rest, albeit with her two children still climbing all over her, she suddenly said to me quite out of the blue, 'I was thinking the other day, if there is no life after death, if it all ends completely when we die, then what's the point of it all? It all just adds up to nothing. The more I think about it, the more it does my head in!'

I have no clever answer to this question, no killer argument, but here is what I think adds some meaning back into life after you have confronted the brevity of life and absorbed the implications of our mortality.

My first thought is to think back over the many fabulous holidays I've been on. On one I trekked through the steamy heat of Mexico to explore the ancient Mayan pyramids in Chichen Itza — a busy stimulating type of holiday. On another I simply lazed around on the white sands of a tropical island in Fiji, swimming and scuba diving between reading books and eating all that fresh tropical food. The point of mentioning those holidays is that they both only lasted for a few weeks before they came to an end and I had to return back home to work. But just because they had to end didn't mean they weren't worthwhile. I didn't say to myself, 'Well, this holiday's going to end so I might as well not go in the first place. After all, what's the point?' If you think about it, the exact opposite is probably true. The very fact that our holidays are so short is what makes them all the more special.

Let's now imagine what happens if life goes on forever — for absolute endless infinity. In this scenario no matter what events unfold, no matter what you do, it just somehow carries on, either on this planet or in some heaven. This infinity of time causes two troubling problems: endless repetition and the fact that nothing is vital.

1. Endless repetition

Given enough time there would be nothing new. There simply couldn't be because after a trillion years what is there left to discover or experience for the first time? And a trillion years is just the very start of infinity. Surely our experiences would become repetitious no matter how wonderful. Think of it this way: imagine playing the perfect game of golf and scoring a hole in one for every shot. As exhilarating as that might be for the first time it must become rather boring after the millionth time.

2. Nothing is vital

Nothing matters because our existence just goes on and on and on. You can't die when you're in Heaven and so there is no consequence for any action. It's just perfect all the time. In contrast a life that's short *is* vital. You have to spend it carefully and your decisions count because they produce different outcomes, some of which are beneficial while others are painful. It's the difference between playing a game where you keep a score and have a winner compared to one which just carries on without any consequence.

So a surplus of time actually degrades life of any meaning rather than adding to life.

When you view a finite life in this way you want to savour each day and spend it wisely. You begin to think of what you can do with the limited time you have available: how you can share happiness with others and add beauty to this world or to the lives of others.

CHAPTER 18
PURPOSE AND MEANING

I COULDN'T FACE LIFE IF THERE WAS NO GOD OR DIVINE PURPOSE

I was chatting to a good friend of mine the other day when she suddenly stopped me, put her hand on my shoulder and looked me straight in the eye: 'You know, what you say about the Universe is probably right but I just don't want to believe it. I couldn't bear it if life was just a random blip on some vast empty cosmic canvas. I choose to believe in a God, even if he doesn't exist, because at least the idea of a God who has a plan for me gives me hope and I can get out of bed each morning thinking life has a purpose.'

This chapter is for Rachel and anyone else who feels daunted by a cold empty Universe without a designer. It doesn't have to be this way and I hope to show that life can be even *more* exciting, valuable and precious without a God than with one.

SLEIGHT OF HAND

We've probably all watched a magician perform some amazing trick that completely fools us into thinking he's done something which simply isn't possible.

No matter how hard you watch you just can't work out how he's done it. Things such as pulling live rabbits out of empty hats or sawing people in half are boring old tricks compared with what magicians get up to these days. Modern 'illusionists' set the bar much higher and produce some truly astounding tricks without any apparent props to support them. These illusions are usually produced by using a technique called 'sleight of hand' where the magician focuses our attention on the wrong thing at precisely the time he is doing something else that's key to the trick. Because our attention is focused on the wrong things, and because he's so slick, we don't notice what actually happened and so we're completely taken in.

It turns out we can also be fooled in exactly the same way by what I call 'linguistic illusions' where a 'flow of words' tricks us into a belief that sounds logical but which turns out to be completely misleading. These linguistic illusions can be every bit as powerful and every bit as convincing as any magic trick. This is especially true when it comes to the question of God, design and meaning in life. Like my friend Rachel, we naturally think that a life designed by a God will have meaning and value whereas a universe without a creator must ultimately be meaningless and sterile. But nothing could be further from the truth.

DENYING THE ANTECEDENT

Like all magic tricks, the best way to see through a linguistic illusion is to slow everything down and make sure we're focused on the right thing at the right time so no hidden assumptions fool us. Let's start by looking at one of the most common errors of language and logic that's directly related to Rachel's concern about life being meaningless without a God.

We begin by *assuming* the following statement is true. It may or may not actually be true but let's just for the moment assume it is:

>   1. *If* 'there is a God who designed the Universe including me' *then* 'life has meaning and value'.

This is quite a reasonable-sounding statement and most people would probably agree with it. After all, if there was a God who made the Universe and had a master plan for everything including me, surely because God is so clever and kind he'll have created my life in such a way it could have value and meaning. All I'd need to do is follow his master plan and ultimately life will be OK.

Where things go wrong is when we take this quite reasonable statement above and automatically assume it implies the opposite statement must also be true.

>   2. *If* 'there was NO God who designed the Universe including me' *then* 'life has NO meaning and value'.

Taking the double negative of our first sentence sounds like quite a sensible thing to do, doesn't it? If you go back and read both sentences (1) and (2) in quick succession one after the other you'll probably find the second sentence seems to follow quite naturally from the first one. And if statement (2) is correct, no wonder people choose to believe in God. After all, who'd bother getting out of bed each morning if life had no meaning or value? I certainly wouldn't. But before we get too depressed let's check to make sure we haven't fallen for an illusion.

To see why we can't jump from the first statement to the second statement we'll use an identical piece of logic, but this time we'll replace the word 'God' with the word 'cat'. Here's how we'll do it. Once again we'll start with a statement everyone can agree is sensible:

    1. *If* '**it is a cat**' *then* '**it will usually have four legs**'.

Now let's see what happens when once again we take the double negative of this statement just as we did in the first example with God. Doing this gives us the following:

    2. *If* '**it is NOT a cat**' *then* '**it will NOT normally have four legs**'.

Aha! We can immediately see the problem. A dog is not a cat but it usually has four legs nevertheless. Clearly, our second statement is wrong and this illustrates an important point. No matter how sensible it sounds to take the double opposite of any true statement — and no matter how convinced you feel in your heart that doing so tells you something of significance — it's going to give you the wrong conclusion. This means that even if God's existence did guarantee life had meaning, we still can't conclude without further evidence that having *no* God would automatically mean life was meaningless. It may or it may not be meaningless; we just don't know yet.

Taking a statement we all agree is quite sensible and then arguing the double opposite must also be true has a long history of fooling people across a wide range of topics. Not surprisingly, this common mistake has been given the fancy title of The Fallacy of Denying the Antecedent. I once even took a British newspaper to the UK Press Complaints Commission over an article where a journalist unwittingly made this mistake in her lead article. I argued that while journalists are allowed to mistake their facts because of our limited knowledge, they must never be allowed to make mistakes in logic. Doing that would be equivalent to bankers deciding they had their own form of Maths where two deposits of $1000 and $1800 only gave you a total balance of $900. You can't do that because Logic and Maths are absolutely true and lie in a special place on our Compass of Truth.

So what have we decided so far?

We've concluded the following.

God $\xrightarrow{\text{implies}}$ **Life has meaning**   **Could be true**

No God $\xrightarrow{\text{implies}}$ **Life has no meaning** **Doesn't always follow**

DESIGN, USE AND VALUE

In the above discussion we've assumed without question that having a creator who designed our life will give it value. But is even this first assumption true? As is often the case, a good way to explore any argument is to depersonalize it and use an analogy. In this case we'll use coal.

Let's just for a moment suppose coal was not designed by God or any higher being but was simply the unplanned by-product of rotting vegetation that became fossilized after being buried for millions of years. It's a waste product that could quite easily have remained hidden without ever being discovered. Now even though we suppose coal was not designed for us and had no innate purpose — because it was just leftover rubbish — one day someone accidentally sets fire to a piece and discovers to their delight it produces masses of long-lasting heat. That is, they found a use for coal despite the fact it wasn't

designed for any original purpose whatsoever. This raises an important question: would coal have any more value to us if it was specifically designed by God to produce heat rather than just being a chance by-product of rotting vegetation? In other words, does adding a designer and giving coal an original purpose increase its usefulness to us? The answer of course is that it doesn't. The heat we obtain from coal is just as valuable whether it was designed for that express purpose or whether it arose by pure chance.

Our coal analogy shows us there isn't an immutable link between an object's original design and its actual value. Another example would be the leather in my shoes. It's just as valuable to me even though a cow might object this wasn't the original purpose of her skin. Moreover, the value I derive from the leather in my

shoes is quite different to the value the cow derives from its skin to keep itself alive and warm. We each derive different values from the same item depending on who is using it and what we each use it for. The only conclusion we can come to from these two examples is that the link should be between use and value and not between design (or designer) and value.

Design ➤ **Value/Meaning**

Use ➤ **Value/Meaning**

But maybe you object to this line of reason by saying: 'Well, that's all very well for objects which we've taken and used for a new unexpected purpose, but what about something like a finely crafted hammer? Surely a well-designed hammer has extra purpose and meaning for the very reason that it was designed?'

Once again we need to slow this argument down and examine in it more detail. We need to ask ourselves: Why is a *designed* hammer more valuable than one that was not designed?

The answer to this question is quite easy. The only reason a designed hammer is better than say a random piece of iron we found lying around is because it's better at hitting nails into wood. Once again this says the value of the hammer is simply related to its usefulness, which is the same conclusion we came to when we talked about coal and cow hide. If there are two identical hammers, one produced by a skilled craftsman and one that somehow magically evolved all by itself, then we'd have to say that because they were absolutely identical they'd each have as much value and use as the other. Their value simply depends on the end result and the use it can be put to, rather than how it arose. Likewise, our value is the same whether we are produced by Evolution or a creator, as long as the final product is the same. This brings us quite naturally to the value of our own lives.

A MEANINGFUL LIFE BASED ON USE

Like coal or a cow's hide, we each have the opportunity of conscripting our own lives for whatever purpose we see fit, and the value we get out of our lives will depend directly on what we choose to do with our lives. A drunken bum who spends his life intoxicated by the side of a road probably won't derive much value from life — whether there is or there isn't a Grand Designer. But life needn't be like this. We each have the opportunity of personally tailoring our own lives. Just as a knife can be used for killing a person or for cutting out a tumour to save a life, so too the value of our lives depends entirely on what we do with it.

This is an exciting and heavy responsibility. We are masters of our own destiny. We need to get off our butts and carefully weigh up our options and put them into action. And isn't something so much more satisfying when you assemble it all by yourself rather than if you just go out and buy it off the shelf? Why wouldn't a tailor-made life we designed by ourselves be any less satisfying than a pre-planned one? I personally think it's more satisfying.

Whether we like it or not, we're already leading lives scripted by ourselves. The only question is: how much care and attention are we each putting into that script? Sure, sometimes we'll be disappointed because our plans don't work out and we'll need to re-script them. Sometimes the random vagaries of life deal us a bad hand, but doesn't this match our everyday observations? Bad things *do* happen to good people. Floods or tsunamis destroy entire villages. Some countries have natural wealth while others are wracked with poverty. And if you think about it for a moment, why would any God choose to curse entire civilizations and design miserable life plans for millions of people based purely on their longitude and latitude when there's a much more logical explanation — rainfall and soil type.

While we can't win every game in life, we're likely to be more successful on average if we use skill and finesse to play each hand we're dealt. This once again reflects our observations. There are winning strategies and there are mental tools that allow us to rise above our natural lot and become more successful. (These tools and strategies are the focus of my previous book *The Winner's Bible*.) It is this scripting and re-scripting of our lives that makes them vibrant and worth living.

> '*It's not the cards you're given, but how well you play the game that counts.*'

THE DECISION TREE OF LIFE

Let's return again to my friend Rachel I mentioned at the start of this chapter. Another reason she wanted God to exist is because this opened up the possibility he'd designed some sort of divine plan or cosmic destiny that was individually tailored for her life. A plan that gave her some sort of assurance her life would be worthwhile if she did the right things and made the right choices. But what does this really mean? Is this really possible? If we dig deep enough we find there are some rather unpleasant surprises hidden inside this concept that we hadn't anticipated. While a divine plan sounds good on the surface it comes at a high price we might not have considered.

Not thinking things through and considering the consequences reminds me of the time I was listening to BBC2 one morning when I was driving in to work. A listener had rung up and was complaining how each morning on his daily train journey into London he passed a group of railway workers standing beside the tracks leaning on their shovels. The caller was outraged they were never doing any work each time he went by. The DJ listened for a moment and then said, 'Have you ever thought they might be working on the track your train was travelling on and they had to stop and get out of the way whenever your train went by?' There was a deathly silence on the phone. The caller hadn't thought it through properly. In the same way, I believe the idea of a cosmic plan for our lives comes with an unexpected cost that is worse than not having a cosmic plan. Let's see why.

We each make millions of choices as we go through life and our choices affect what happens to us and our happiness. Some of our choices have major significance, such as which country we live in, who we marry and what type of job we do, while other choices have little impact, such as what we're going to eat for dinner tonight. Nevertheless, each choice we take opens up thousands of new paths and simultaneously closes down millions of others. If you move to Hong Kong to take up a new job then you won't be able to marry the person who'll visit your old favourite pub on the spur of the moment next month.

We can represent all your life's choices and the paths they open up or close down by using a tree diagram like the one below.

To use the diagram, we start at the bottom of the tree and move upward in time to the next choice you have to make in your life, which is represented by two diverging branches, A and B. For whatever reason, at this moment in time you are faced with a decision which requires you to choose one option or the other — either A or B. (For simplicity we've only given each choice two outcomes but it wouldn't matter if there were 10 or

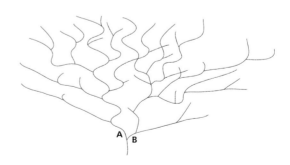

100 options at each decision point in your life. More choices would simply mean more branches and a messier diagram.) Regardless of what your decision is about, whether it's which clothes you're going to wear tonight or whether you're going to have another glass of beer before you drive into town, the decision you settle on opens up some possibilities and closes down others.

This closing down of the options you didn't take is why the branches diverge at each decision point in your life. As you move along the branch you've just chosen it's not long before you're faced with a new decision which is represented by another fork splitting into two more branches. Before long your tree of life has branched out into a myriad of diverse possibilities, most of which you didn't take. We can represent the path you did take by colouring it in red in the diagram below. Every other branch you didn't take will forever be hidden from you. You'll never know how they would have turned out or what consequences or opportunities would have ensued if you had taken them instead.

Some of your decisions will seem trivial at the time but they may have totally unexpected consequences. For example, you might be out for a ride on your motorbike one day and decide on the spur of the moment to stop at the nearest café for a quick cup of coffee. No sooner have 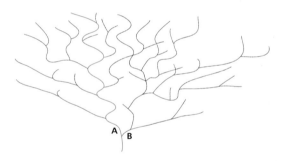 you settled down by the window to drink your latte than another biker rides up with a similar intention. Before long the two of you have struck up a conversation kindled by your obvious common interest in bikes, not knowing it will ultimately grow into a lifelong friendship. Even more unexpected was that you'll meet your future wife in three years' time at a BBQ at his place. A simple, apparently insignificant cup of coffee at a randomly chosen café changed your entire destiny — not to mention that of your future wife and your children.

In a strange sort of way, the path you actually follow on this tree fully describes your life. It summarizes all the relationships, experiences, successes, failures, disappointments, happiness and sadness you'll have, because each path contains its own unique set of events. In a way it encapsulates your life in a tight, compact manner.

But what does all this have to do with the concept of a Divine Plan or Cosmic Destiny?

An awful lot, as it turns out.

WHAT DOES A DIVINE PLAN OR COSMIC DESTINY REALLY MEAN?

We can use our tree diagram to explore what it would mean if there really was Cosmic Destiny or a Divine Plan for each of us. However, before we can do this we must at least define what we mean by those words Divine Plan. The usual definition is that there's some optimal or preferred path that is best for us. We can represent this 'best path' by colouring one of the paths red so it stands out from all the other paths which aren't best for us.

Notice we haven't yet decided what this concept of 'best' means. It might be measured purely in terms of our happiness here on Earth or it could be measured by taking a much wider view where we need to include the concept of an afterlife. It might be that our mortal lives are little more than preparation for eternity and so a life of misery and suffering on Earth purifying our souls could easily be outweighed by spending eternity in Heaven with God. Likewise, our human measures of 'best' could be completely wrong because we might base them on our transient happiness, our contentment or success, whereas a divine agent might weigh the different paths using an entirely different scale. A divine measure of 'best' might focus on how advanced our souls become rather than how happy we are.

Now even though we can't decide which path is 'best' because of our ant-like limitations, all we need to agree is that *if* there is some divine plan or cosmic destiny for us, there must be some path that our divine agent has mapped out for us. There must be one red line that's better than all the other less worthwhile options. This is exactly what my friend Rachel wanted at the start of this chapter. Some sort of assurance that if she 'did the right things' then her life would somehow work out. That there was a 'best' path in red personally tailored for her.

But is this really possible??

THE PROBLEM OF FREE WILL

The immediate problem we face with this idea is: how can any divine plan be reconciled with the notion of personal free will? To see why there might be a problem we need to zoom in to one of the branches on our tree at a point where we're just about to make a choice. In the diagram on the right we've zoomed in really tight so we can't see all the millions of choices we've previously made in the past or all the choices that'll be available to us in the future. Instead we've focused entirely on the very next choice we're going to make. In the next instant we'll either go down option A or we'll go down option B.

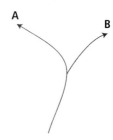

To make the argument clear let's illustrate these two choices by taking an extreme example. Suppose you're a college superstar and you've just received two letters. One invites you to attend Harvard Medical School to become an eye surgeon (option A) and the other is a contract for you to become a quarterback with the San Francisco 49ers (option B). Two fantastic career opportunities, but obviously you can't do both at the same time. After much deliberation you decide to take the Harvard offer and so there you are, all dressed up, driving to meet the Dean to personally accept his offer. But just as you're about to put this choice into action a girl you've never met is going to make a decision that might ruin your life. She's been to a party where she's drunk too much and now has a choice of whether to drive home from the party or to catch a taxi. We can represent her two choices, drive home (option C) or catch a taxi (option D), by the two green lines in the next diagram. Unfortunately, she decides to drive (option C) with the catastrophic consequences that she crashes into your car, causing you serious injuries. You're left with permanent, total blindness and are paralyzed for life. Both your previous options of being an eye surgeon or a football star have gone. Instead, your life now moves off on a new line — path C.

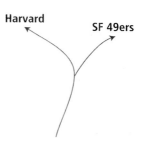

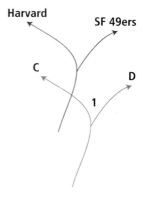

This brings us clearly to the nub of the problem. The drunken woman had complete free choice at point 1 — she could either have driven home or not. If she hadn't driven you would still have those options of Harvard or football open to you and they might have been paths on your red line — your ultimate 'best destiny'. But now they're gone forever and there is no guarantee that any line branching out from option C is going to be as good, or even any good at all. This brings us face to face with the impossibility of reconciling free will with a cosmic plan.

EITHER WAY YOU'RE STUCK

To see why there's such a problem with free will and a cosmic destiny, let's just for a moment assume you're on your red line representing the plan that leads you towards your destiny. This is shown in the diagram below in the first panel labelled 1. Once again, someone else is about to make a choice that is going to affect your life. This is shown in the second panel.

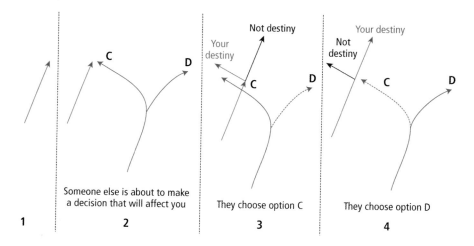

If they choose option C (for example, they might accidentally blind you) then your life must change direction because their choice C interferes with your life (Panel 3), but if they choose option D then your life will carry on as it was before (Panel 4). Now here's the catch ... whatever they choose, your life must still be on the red best line of your perfect plan taking you towards your destiny. But this means you'll get two different divine plans depending on their choice. In one your red line goes left and in the other your red line goes straight on up.

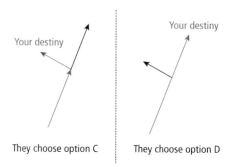

This is impossible because these two completely different paths can't both be the *best* plan at the same time!

GOD CAN MAKE IT WORK OUT AFTER ALL

One way around this dilemma is for us to say, 'Well, no matter what anyone else does, God can turn any option (including C where I'm blinded) into the best one.'

This is equivalent to saying two things:

1. It ultimately doesn't matter what anyone else does to us — it will still work out.
2. There are an infinite number of 'best paths'.

But these two statements come with unexpected fishhooks.

If we can lead equally optimal lives no matter what anyone does then there can be no concept of 'right and wrong' and no penalties. After all, if someone kills my wife or steals all my money and this still leaves me on a path which produces the exact same amount of 'bestness' in my life and which still takes me to the same destiny, then why should we punish anyone? According to this argument no one is any worse off because God can always find a way around whatever you do to anyone. If this is true then it's hard to see why there would be any reason for a hell to exist because there are absolutely *no* negative consequences of any actions. Using this line of argument we should be just as happy when someone gets cancer or becomes maimed in a car accident as we would if they won the lottery or a gold medal at the Olympics because all these paths lead to the same amount of 'bestness'.

This argument about 'God turning any option into the best one' isn't just confined to bad things either. It also means we needn't try to do anything good because whatever we do, the total happiness in the world won't change for the reason that whatever we do isn't allowed to affect the 'best outcome'. Surely, this argument just doesn't make any sense. It goes against the entire essence of what it means to be a human: to fight for right and wrong, to innovate and find cures for diseases such as polio which destroyed millions of lives.

Secondly, if God can 'always make things work out' then there must be an infinite number of these 'best paths' because the number of ways different people can affect us is limitless. Think of it for a moment. You invest your life savings and take out a sensible mortgage to buy a new house for your family in New Zealand. Everyone says you got a bargain and it's easily affordable on your salary. A great investment for the future. But then, someone at Lehman Brothers makes a decision which ultimately sends that great company into bankruptcy which then triggers the global financial meltdown of 2008/2009. That in turn reaches all the way down to New Zealand a year later where you lose your job and can no longer afford your mortgage. Next thing you know your house is foreclosed in a severely depressed market and you now have negative equity. Your family is on the streets with zero dollars to your name. Small changes on the other side of the world affect you in complex ways you cannot even fathom. All these billions of choices made by millions of people indirectly affect your life; yet somehow, magically, each of these billion different paths open or closed to you — depending on what other people choose to do — must all balance out and produce the exact same net outcome. For if one path was even slightly better than the other, you'd have to take that one and all the others would immediately cease to be red-line options. But this means no one else can do anything to affect you.

Adding in the concept of an afterlife adds nothing to this argument whatsoever unless this life on Earth counts for absolutely *zero* in the calculation of 'bestness'. But that in turn means our mortal lives are totally and utterly meaningless. The exact opposite conclusion to what Rachel was after in the beginning.

WE DON'T HAVE FREE CHOICE AFTER ALL

Another way around this dilemma is to assume that our drunken lady is somehow prevented from making any decision which affects you. In other words people either:

1. Don't have any free choice at all.

    or

2. They have free choice but in such a limited manner they can never interfere with God's plan for you.

If we take the first option and say our woman is not 'free' to make her choice and is somehow pre-programmed to do certain things then we immediately remove any possibility she can be held accountable for her actions. After all, someone else is controlling her life and 'made her do it'. We've all become nothing more than puppets with someone else pulling our strings. Once again we end up with the situation where there can be no moral responsibility and no Heaven or Hell.

The second option, of having free choice which is contained in some sort of protective bubble that isolates us from causing any harm to anyone else, also hits a brick wall of impossibility. If everything we do in our 'bubble' is prevented from affecting the destiny or happiness of anyone else in the entire Universe then once again we're forced to conclude there is no reason to punish or reward anyone. The only damage we can do is to ourselves, trapped in our own isolated bubble. But even this isn't enough of a restriction. We notice people *do* come into contact with our lives on a regular basis and if these people genuinely have free will then they must have the option of interfering with our life in a negative manner.

So what are we to conclude?

WHAT DO WE ACTUALLY OBSERVE IN REALITY?

If our lives have any value and if we are responsible for our actions and can influence the lives of other people — which is surely what we observe in everyday life — then we must conclude our free will carries with it a heavy price, namely there are *no* guarantees we'll ever lead the 'best life'. All we can ever hope to do is play the game of life as best we can and hope that on average we are given an 'even break'. It's this uncertainty which makes life both raw and ultimately worth living. After all, if you knew the outcome of every tennis match in advance it would hardly have much attraction. It's this very uncertainty which makes our decisions worth worrying over because they carry so much weight. It's why we need to choose our friends and associates so carefully, why we need to plan for the unexpected contingencies.

Life is harsh and only the very spoilt think otherwise. Actions of people such as Hitler do cause tremendous damage and suffering. Surely, the best of all worlds is when we realize life is a dangerous adventure with no guarantees but, regardless, we play the game of life with wisdom and optimism, eagerly awaiting the next unpredictable and exciting outcome.

FINDING GOOD IN EVERY SITUATION

In closing, there is still one more aspect we need to cover. Part of the power of believing in a divine plan or a cosmic destiny is that we always remain positive regardless of what happens to us. We continually look on the bright side and try to make the best of any situation because we have a belief that somehow it will work out for the best. As I discussed in Chapter 12, The Psychology of Belief, the power of having such a positive mental attitude is inestimable, whether it's inspired by a fictional supernatural force or generated from within by our own strength of character. If we believe there's always a positive future ahead and if we keep on going and make wise choices with our heads held high, it will more than likely come to pass. Not because of any cosmic force but because we have put ourselves in the optimal mental state for success.

> 'The ideals which have lighted my way, and time after time have given me new courage to face life cheerfully, have been kindness, beauty and truth.'
> — Albert Einstein

CHAPTER 19
ETHICS

'The most important human endeavour is the striving for morality in our actions. Our inner balance and even our very existence depend on it. Only morality in our actions can give beauty and dignity to life.'
— Albert Einstein

BEYOND BELIEF

Twenty-year-old Ashraf Haziq was a gifted Malaysian student who'd recently won an accountancy scholarship to study in England. As usual, he was quietly riding his bicycle through Hackney in the early summer sun to meet some friends for breakfast. Within minutes of leaving home Ashraf was violently attacked by a group of thugs who knocked out his teeth and broke his jaw so badly he later needed to have a steel plate inserted in his jaw. What was particularly shocking about this violent and sustained attack in broad daylight was that most of the attackers were aged around 13 years old. Their motive? The barbaric 'pleasure' of inflicting serious injury on someone else. It was as simple as that.

Once Ashraf had been beaten to the ground there seemed little else for these juvenile thugs to do so they stole his bicycle as some sort of trophy they could later display to their friends. For the next four minutes onlookers simply walked past Ashraf, ignoring his distress as he lay in a pool of his  own blood. Finally, two older youths came over and helped him to his feet in an apparent act of concern. But instead of helping him, one of the youths proceeded to rifle through his knapsack and steal his wallet, phone and Sony PlayStation while the other restrained him. The entire event was caught on video as you can see on the YouTube video clip (http://www.youtube.com/watch?v=6Gex_ya4-Oo).

How anyone with even the slightest vestige of humanity could inflict such a senseless beating on a completely innocent person is hard to imagine. How a second group could then look closely at his horrific injuries and proceed to steal his possessions instead of having compassion is simply beyond belief.

This example is just one of many thousands that occurred during the London riots of August 2011. The only thing that is special about this crime was that it was caught on video, allowing us to see the painful human cost of what would otherwise be a simple, lonely statistic: Youth Assaulted.

Unfortunately, events like this are not unique to England. Los Angeles experienced riots on a similar scale to those in London immediately after the Rodney King court case. During six days of riots 3600 fires were lit and 1000 buildings were burnt to the ground. Innocent bystanders were shot for no reason and there was widespread looting with thieves brazenly brandishing their spoils to TV news crews. Law and order was only restored after the White House ordered the military to lock down the streets. By then 60 people had been killed and more than 10,000 arrested. And the list goes on and on. Philadelphia has been forced to impose nightly curfews on all teenagers to stop widespread violence and looting. And if we take a look at the big picture we see that serious crime rates across most Western countries have increased by *six* times over the last 50 years. That's not the total amount of crime, but the amount of crime committed by each individual!

So what on Earth is happening to our so-called civilizations and what can we do about it? Has Western society somehow changed beneath our noses without us realizing it?

THE MORAL SCALE

Without doubt there were many causes of these riots and I'm sure various government agencies, commissions and social welfare groups will do their very best over the following months and years to come up with solutions. While deprived social conditions and poverty are likely to be contributing factors, they can't be the key underlying causes because many of those arrested in both riots were actually in gainful employment, including a 20-year-old school teacher who was one of the first arrested in London. And it can't simply be a lack of policing or social welfare because both of those are at record highs.

So what is it?

My approach is to start from a rather unconventional position and begin by trying to identify a single ethical problem which is at the base of all crime. If we can indeed find this root cause and if we can find a way to measure its size in each individual then:

- we can use this to shape our future policies
- we can determine the effectiveness of policies after they are implemented by determining how many individuals have changed and by how much.

Ideally, our measurement should be as simple as possible. Here's one idea I think has merit.

Just as we can measure each person by their height or their weight, their IQ or their wealth, so too we could measure each person on a Moral Scale. In this scheme a newborn baby starts out at Level 1 and is concerned only with itself. It cries when it's hungry and it has no concept of the needs of others. We can represent this undeveloped natural state by the small circle at the very bottom of the diagram.

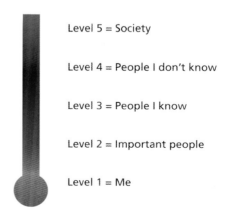

The Moral Scale

Level 5 = Society

Level 4 = People I don't know

Level 3 = People I know

Level 2 = Important people

Level 1 = Me

Over time the child learns that other children have needs just as they do and so they learn to share the toy they selfishly wanted to play with all the time. Initially, this is enforced by an adult through a sense of 'fair play', but as they grow up, they become genuinely aware of other people's feelings and develop a sense of what it must be like to be in the other person's shoes. And so they move on to Level 2. At this still primitive level of development they don't do things that would harm people who are important to them.

They're careful not to harm their friends because their friends add value to their lives. In the same way, they probably wouldn't steal from a famous celebrity they admired, like Raphael Nadal, even if they don't personally know him, because in a roundabout way that celebrity has some importance for them. Level 2 is still a completely selfish level of moral development because any violation at this level is likely to cause painful repercussions, such as losing friends or the respect of people they care about.

By the time a person has progressed to Level 3 they are genuinely concerned for the welfare of people they have met even if those people aren't important to them. These people could be someone they meet on a street or a person who goes to the same gym as they do. Because they've looked into their eyes and seen there is a beating heart and soul attached to that person their compassion is aroused, even if these strangers aren't personally important to a Level 3 person.

Quite clearly the children who assaulted Ashraf Haziq hadn't progressed past Level 2 because they could look into his eyes, and then with callous disregard for his plight, knock out his teeth and break his jaw. It was a graphic demonstration they didn't care about other humans per se but were instead ruled entirely by selfishness. In other words, they were Level 2 humans that had progressed little beyond a crying newborn baby of Level 1 and equal to a two-year-old kindergarten child. This illustrates just how primitive the first levels of the Moral Scale are.

At Level 4 we become genuinely concerned about *all* people, including those we've never met. We don't need to be reminded by looking into someone's face that every person has feelings like we do. On finding a wallet dropped in a street, a Level 4 person would look at the driver's licence inside the wallet and try to find the owner so they could return the wallet to them. Even though it might cost time and money to do so, a Level 4 person understands how losing a wallet affects an abstract individual. They don't need to have met that person to know the wallet is attached to a set of emotions, dreams and desires of a real person. In a similar way, a Level 4 person would never shoplift because even though that iPod on the shelf is not directly attached to a person, it *is* connected to the shareholders or owners of the store who are part of humanity. They wouldn't want someone to steal from their shop if they owned one and so they won't steal either.

I think by now it is pretty obvious that as we move up through the levels we quickly restrict the types of crime a person in each level could commit. By the time we reach Level 4 we have effectively eliminated all crime. Regardless of your religion, of any laws or what you have been taught is right or wrong, a person who has genuinely developed to Level 4 is so acutely aware of their actions on the wider community and other people that it is effectively impossible for them to commit a crime.

But there is still one higher level to go.

> *'The destiny of civilized humanity depends more than ever on the moral forces it is capable of generating.'* — Albert Einstein

At the highest level, Level 5, a person is now concerned with the entire fabric of society and not just with the individuals that make it up. They are concerned with the thoughts and beliefs that pervade society, the goals of society and the unwritten rules by which it operates. They debate questions such as whether we should allocate more of our precious resources on infrastructure or whether we should spend it instead on health initiatives. Their interests range from policy on unemployment to the content of video games available to young impressionable minds. They care about the big picture and all the messy political issues because they are aware these questions intimately affect every person in society; from those at the bottom to those at the top. After all, a society governed by the rules of communism is likely to provide different outcomes for the population than a society governed by capitalism. A Level 5 person not only weighs these issues up but is also pro-active in ensuring the best ideas are put into practice and become effective.

ANCIENT GREECE AND ROME

If you read the writings of Socrates, Plato and Aristotle, to name but three of the great Greek philosophers, you will see they were passionately concerned with Level 5. Unlike modern academic philosophers at universities they also made sure their thoughts influenced the wider society and were absorbed by the common citizen. In the case of Socrates, he was even willing to die for his political and sociological beliefs. It's no wonder these ancient civilizations flourished and produced some of the most remarkable advances in humankind's history. Think about it for a moment: they were able to build huge libraries of great knowledge, provide water-reticulation systems to entire cities and calculate the distance to the moon, the radius of the Earth and the size of the sun 2500 years ago.

They were advanced because their society had a well-defined goal and each citizen understood their role in achieving the common objective. There was a sense of belonging to a proud and noble society. Nowhere was this more evident than at the height of the great Roman Empire. To be a Roman citizen in those days was something to be incredibly proud of. Wherever the initials SPQR (Senatus Populusque Romanus — the Senate and People of Rome) appeared, whether carved on a building or a military standard, you knew you were at the centre of civilization.

Now contrast this to the youths who ransacked LA, London or New Orleans. During TV interviews it was clear they resented the society they lived in and rather than wanting to contribute to it, they were actively seeking to damage it. I can't imagine any of these youths even remotely relating to a concept like the 'common good of the United Kingdom', something we might call SPQUK. The nearest they get to supporting anything seems to be their local football team.

The power of a country having a commonality of purpose is not confined to the rose-tinted history of ancient Greece or Rome. We saw similar effects when America united in a 'race to the moon'. This led to a dramatic spurt in technical innovation which flowed beyond the gates of NASA and increased the standard of living of the average American. Within short order homes became populated with all the modern electronic conveniences we now take for granted.

COST OF CRIME

Let's pause for a moment to consider the financial cost of crime caused by people with low levels on our Moral Scale. It has been estimated the annual total cost of crime in the UK is around £64 billion when one takes into account the damage done to individuals, the cost of police, courts, jails and so on. In the US a comprehensive economic analysis put the figure at $1.7 trillion annually (US$1,700,000,000,000). These are truly staggering numbers but they only tell part of the story. (See: http://www2.davidson.edu/news/news_archives/ archives99/9910_anderson.html)

In the UK there are approximately 8000 violent crimes and 5000 house burglaries every day! What constitutes rape and sexual assault and what percentage of occurrences is actually reported varies greatly from country to country. However, in the US, for example, there are 300 (male–female) rapes *reported* every single day of the year. Think about it. On Monday 300 women had their lives devastated by a rape. On Tuesday another 300 women's lives were devastated … and so on. One every five minutes — morning and night. The sheer scale of crime behind these numbers beggars belief.

THE PERSONAL IMPACT OF CRIME ON YOU

What's not obvious when you read statistics like these is that each person in society, including you, is affected by every crime that is committed. Even if a crime happens thousands of kilometres away from where you live, it still touches your life because the cost of every crime must come from somewhere. Imagine how *your* life would be transformed if the government suddenly had an extra US$1700 billion to spend *every* year on civic amenities such as swimming pools, sports stadia, parks, education, healthcare, the arts and so on. It wouldn't take long before everyone enjoyed the sort of life currently only available to multi-millionaires. Even consumer goods and luxury items like cars and yachts made by private companies would be cheaper because sales tax and personal income tax could be slashed to a fraction of their current levels if the government didn't have to fund the cost of crime. And that's not even taking into account the savings on things like insurance. If we could make a serious dent in crime then with all the modern technology available to us we'd live truly remarkable lives that were unprecedented in human history.

THE BIGGEST RETURN ON INVESTMENT

It's hard to imagine *any* investment giving a greater return to society than one that raises the average level of morality in society. No new wonder-drug, electronic invention, housing project or infrastructure programme would provide even 1 per cent of the benefit that would accrue to society if every citizen operated at Level 4. Even winning the 'war on terror' pales into insignificance beside the 'war on morals' because American citizens are at a far higher risk of being killed by their friends and neighbours than they are by Al-Qaeda. And yet the US Government has spent $1.27 trillion on the 'war on terror' but the corresponding 'war on morals' is to all intents and purposes unfunded, despite the rewards being vastly greater and the probability of success being higher.

THE WIDER ISSUE

Before we look at the practical steps we can take to improve society's average 'moral score' I'd like to talk about the wider impact of having Level 1 to 3 citizens loose in society. So far we've only talked about 'hard' crimes like murders, burglaries and rapes, but there is a wider issue where selfish antisocial behaviour reduces our quality of life even if a crime is not committed. There are plenty of examples: people playing their stereo at full volume in the middle of the night without any regard for their neighbours trying to sleep at the end of a busy day. Cigarette butts and fast-food wrappers flung from car windows because the occupants couldn't be bothered to put them in a rubbish bin. Inconsiderate drivers … the list goes on and on. You don't need me to list them all because you've probably got your own list of pet antisocial behaviours that annoy you.

ETHICS AND RELIGION

'The greatest tragedy of mankind may be the hijacking of morality by religion.' — Arthur C. Clarke

The five Levels of Morality all rest on one fundamental principle, which is often referred to as the Golden Rule: **'Do unto others as you would have them do unto you.'**

Our Moral Scale simply shows how far and to whom a person is willing to apply that rule. Do they just apply it to their friends, to people they have met, or all the way to include that abstract notion of the faceless, impersonal society? While many think the Golden Rule is a product of religions like Christianity or Islam, it actually predates those religions by many generations as the following quotes show:

- *'Do to others as thou wouldst they should do to thee, and do to none other but as thou wouldst be done to.'* — Socrates

- *'May I do to others as I would that they should do unto me.'* — Plato

- *'Do not impose on others what you yourself do not desire.'* — Confucius, Doctrine of the Mean 13.3

- *'Do not wrong or hate your neighbour. For it is not he who you wrong, but yourself.'* — Pima (Native American) proverb

- *'Hurt not others in ways that you yourself would find hurtful.'* — Sakyamuni Buddha, Udana-Varga 5:18 (Buddhism)

It turns out those religions borrowed or adopted the Golden Rule rather than the other way around. But this adoption doesn't come without an immense social cost as the quote above from Arthur C. Clarke claims. Let's consider just two reasons why allowing Religion to be involved in the development of our citizens' ethical and moral fibre is such a costly mistake.

1. When belief in God fails

Suppose we are told 'all morals originate from God or by divine decree' — as most religions claim. This works brilliantly while you believe in God and the threat of an eternal hell to punish anyone who transgresses. But what happens when a person or whole tranches of society lose their religious belief under this scenario? If the *only* reason they've been taught to behave ethically is because of God or the threat of Hell, then when they lose their belief in God they also lose their reason to behave correctly. Suddenly, it's 'everyone for themselves'.

This is particularly disastrous when any society transitions from being one that was predominantly religious to one that is increasingly 'post religious'. During the change-over you suddenly end up with huge numbers of people who have absolutely no basis for their ethics. Of course some people will spontaneously go on to develop their own ethical underpinning, but unfortunately many don't without external guidance. In fact sociology shows that during these transition periods the pendulum often swings too far and the vast majority follow along like sheep.

2. Barrier to social ethics

A second problem arises when religions claim they're the one true source of moral codes. When this happens the religious authorities either lobby against secular Ethics being taught at schools or they crowd out the curriculum with their own teaching to ensure there is no room for alternative Ethics. If neither of those routes works then they insist the issue of ethics is so contentious that no one in public office should be allowed to teach it. We then end up with the double whammy where schools are unable to teach secular Ethics because of the political power of the Church while at the same time a significant percentage of the population doesn't attend a church either. This is the worst of both worlds. The net result is a moral vacuum where Ethics isn't taught anywhere.

Matters quickly spiral downhill because the most powerful forces left to shape young minds become the voices of consumerism and the adverts we are bombarded with each day. Unfortunately, while capitalism is staggeringly successful, at its heart it embraces a 'self-centred ethos of competition' and 'everyone for themselves'. If I build a better mousetrap and I can beat my competitor then I'll make more money. And the more I can grab the better I am (see Chapter 20: Society).

This already dire situation is compounded because of the content of television. By the time the average American child reaches the age of eight they will have seen 8000 murders and witnessed 100,000 acts of violence on TV. These frightening statistics are made even worse because the pernicious morals on these shows are not balanced by parental teaching or guidance. According to a recent Nielsen survey the average American child aged two to five years now spends approximately 32 hours a week watching TV, while for children 12 years old this grows to 53 hours per week when video games and the internet are factored in. At the same time studies show the amount of quality time working parents spend with their children is around 25 minutes per day.

CONTRIBUTING FACTORS

Before we turn our attention to the practical steps we can take to increase society's Moral Score, it's probably worth commenting on a few factors that have contributed to this current state of affairs. None of these factors was sufficient on their own to have caused the decline but taken together they have conspired to produce a perfect storm of moral decay.

Social media

It is a growing trend for people to now have more 'social interactions' and 'friends' online than they do face to face in real life. Facebook, Twitter, blogs and innumerable chat rooms have produced an explosion of online socializing that has quickly overtaken the good old-fashioned meeting. But the problem with these virtual relationships, as you probably know from your own emails, is that it's much easier to be harsh or downright rude to someone in cyberspace than it is while standing directly in front of them. It's easier to say something nasty when they can't lash out and punch you on the nose or you can't see the pain in their eyes your nasty comments have caused. It's both tempting and easier to treat people in cyberspace less considerately than people you meet in person. And if the majority of your relationships are virtual then over time you will become desensitized to the feelings of *real* people.

The net result of all this is you begin to slide down the Moral Scale because other people's well-being becomes less important to you.

My rights versus my responsibility

Another relatively recent phenomenon is the increased emphasis on 'individual rights'. Instead of championing the common cause everyone today seems to be claiming their right to do whatever they like and the freedom to say whatever they want. But we seem to have forgotten that 'rights' come with 'responsibilities'. If we have the 'right' to drive a car then surely we have a 'responsibility' to not hurt anyone else on the road. If we have the 'right' to freedom of speech then we should be responsible for any harm our words cause. This disassociation of 'rights' from 'responsibilities' again reduces the connection between our own personal actions and their effects on other people. And being concerned about our actions on other people's well-being is exactly what the Moral Scale is all about.

Capitalistic emphasis

As you will see in Chapter 20 (Society), the capitalist/free enterprise/consumerist system has been staggeringly successful in the Western world for a number of centuries now. But unless the tenets of 'competition' and 'winning at someone else's expense' are balanced with an equally strong sense of social responsibility, we end up with a 'selfish me' generation. What can I get? How can I win? What's in it for me? Once again this sort of outlook reduces our concern for our fellow citizen, which has the net effect of gradually reducing our level on the Moral Scale.

Lack of consequences

This is a rather unusual one because in some ways it seems to work in the opposite direction to the capitalist problem listed above while still managing to cause the same negative outcomes. In this case people don't take responsibility for their own actions. If they smoke and get lung cancer, they expect society to provide free healthcare to treat them. If they are overweight, eat junk food and don't exercise, they still expect the same treatment and care as someone else who has done all the right things and led a healthy life. If they don't study or work hard or make risky choices in life they expect the government to bail them out when things go wrong. What produces a particularly toxic mix is when you have a 'Me generation' person who wants everything for themselves but who thinks at the same time that society (everyone else) needs to look after them. That produces the worst of all possible worlds.

As I said at the beginning of this short section, these are only contributing factors and not the causes of moral decay, but they definitely don't help and we need to be aware of them as we consider some ideas to tackle moral decay.

INITIATIVE 1: LESSONS FROM THE GLOBAL WARMING DEBATE

'The broken society is back at the top of my agenda.' — British Prime Minister David Cameron

What is interesting about this chapter on ethics and crime in society is that I wrote it back in 2005, long before the riots in London and long before 'the broken society' was recognized by any political party in the UK as being an important topic. I've simply updated it with contemporary events to coincide with the release of this book. To me, the dissolution of society's ethics was always a ticking time-bomb.

What concerns me is that politicians still don't seem to have understood the root cause of the problem. Instead there are loud calls from Parliament for increasing police numbers, for harsher penalties for crime and a crackdown on gangs. While those are worthwhile initiatives they still don't get at the heart of the problem. We're already living in the most incarcerated society in history and stronger punishment and deterrents clearly aren't the complete solution. Despite ever-increasing laws our crime rates have sky-rocketed in the last 50 years to six times the rate we previously experienced. Adding more legislation and punishment is the equivalent of putting more ambulances at the bottom of the cliff to treat those who've fallen off while the crowds at the top of the cliff continue to grow. We need solutions that change the individual from the inside out so they no longer want to commit crime in the first place. And that gets back to the moral level of each individual.

Let me start with a fairly radical suggestion and use global warming as a case study. For the purposes of this argument it doesn't matter whether you believe in global warming or not; what we're interested in is how that debate progressed and the lessons we can learn from it.

For years scientists had been publishing scholarly articles warning of the perils of global warming and it made absolutely no difference whatsoever. The average person went about their lives completely unaffected. Then the contentious movie *An Inconvenient Truth* was released. This painted in stark everyday language the consequences of global warming and to some extent kick-started the public debate. But, still, government policies remained resolutely unaffected and the net result was little more than an extremely vigorous debate with people taking both sides of the argument.

The turning point came when a paper was published which clearly spelt out the financial cost of global warming to every government and to each individual if nothing was done. Money! That's the sort of language people and governments respond to. Before long all sorts of initiatives to prevent global warming were instituted, including carbon tax and carbon credits. In the same way as with global warming, we need to show the genuine financial cost each and every individual in society pays for moral decay — from young teenagers to retired pensioners. This needs to be done in a graphic way where we use the same sort of emotional images as the global warming activists used when they showed polar bears clinging to melting icebergs. TV adverts showing images of young children going to school without proper food or clothing because we've spent their money on crime. Pensioners huddled under blankets trying to keep warm because we couldn't subsidize the heating for retirement villages. A young mother dying because the health system ran out of money. An empty barren plot of land that should house a sports recreation facility. With images like this the true cost of crime would quickly jump to the top of the agenda so it became part of every economic policy and treasury discussion.

The credit/tax system

Next we need to put in place a model which rewards any enterprise that advances society or individuals up the Moral Scale and taxes those the hinder it. Just as we put a 'gas-guzzler' tax on cars that burn too much fuel or on coal-fired power stations because of their emissions, so too we should tax enterprises that produce items like violent video games or TV shows that numb individuals to the consequences of their actions. Conversely, any organization that advances morality should be given tax breaks or credits.

Let's take a tangible example. If the huge marketing giant Nike, with all its sponsored celebrities, decided to include a genuinely positive social message in all its advertising that was sociologically proven to influence young impressionable minds, we should give them the equivalent of a 'carbon credit' and reduce their taxes. Maybe they could discreetly embroider a simple slogan like 'Think of the other fellow' on their products or maybe Le Bron James could use this as his closing line whenever he endorses a product. The acronym TOTOF might even begin to seep into the public consciousness if it was used enough.

There is no question these powerful organizations exert a huge influence over society as witnessed by the massive premium impoverished youths are willing to pay to 'wear the right brand'. Did you notice how many of the rioters were wearing expensive brands like Nike, Adidas and so on? Almost every single one of them!

Of course estimating the social value or cost is not going to be an easy thing to do but neither is it easy working out the true carbon cost of dairy farming or of running a power plant. But with the right willpower we managed to do a pretty good job anyway. And sure, there'll be many companies that moan and resist fronting up for their share of social responsibility, but that's no different to the auto industry moaning about gas-guzzler tax. Even if we make some mistakes in our tax calculations it's a far better long-term solution than rubber bullets and tear gas to control a society that simply doesn't want to be controlled.

This simple germ of an idea when fully worked out could have profound influences on all aspects of society ranging from TV programming all the way through to how our celebrities behave. For example, any celebrity engaged in antisocial or 'unsocial' behaviour should immediately have all their PR and media oxygen shut off. After all, they are only celebrities because of this exposure in the first place. Doing this will require strong incentives for tabloids and news organizations which seem to relish documenting bad behaviour rather than reporting on good deeds.

INITIATIVE 2: EDUCATION

The second strongest institutional influence, following behind advertising and commerce, is our schools. We need to make sure a key measurable outcome from our schooling system is that every single individual is fit for society. There is absolutely no point in teaching a child Pythagoras if they haven't learnt how to be a positive and productive part of society. This will entail far more than just giving them a set of 'rules' or behavioural guidelines they ought to follow.

Rules like that just don't work; in fact they often achieve the exact opposite. Instead, the key will be to create an environment where it is 'cool' to be ethically developed and uncool to be ethically immature. Where each student's 'attractiveness' and acceptance by their peers is related to their level of moral maturity. While this might sound like a difficult task, it is exactly the sort of outcome that advertising and marketing campaigns regularly achieve. If you wear Nike you are like Michael Jordan and cool. If you drink Red Bull you're an adventurous extreme sport nut.

So rather than having morals and ethics taught as part of religious studies they need to be entwined in a secular curriculum as if schools were engaged in a marketing campaign to change people's preferences.

An initiative like this should be wrapped inside a process that increases each person's awareness and appreciation of beauty rather than simply encouraging them to accumulate more material goods. We know this process works because it has worked in every civilization where it has been a priority — from the ancient Greek civilizations through the ages of Enlightenment to modern times.

As a person who has previously been involved in setting up training programmes for world champions from sports such as Formula 1 and the Olympics, I am a passionate believer in what I call objective metrics. This involves a three-step process:

1. Identify what goes into making an elite performance

First, you need to identify all the components and subcomponents that go into making elite performance. For example, for a tennis player you might start off at the top level with items such as the athlete's physical ability, their stroke making, mental ability, tactical decision making and so on.

2. Break the performance into subcomponents

Next, under each of these high-level items you break their performance down to the next level subcomponents. For example, with their stroke making you now consider their serve, forehand, backhand, volley, slice and so on. For their physical ability you might consider their aerobic fitness, endurance, ability to cope with heat stress, speed, agility, flexibility, injuries and so on. And then once more you drill down to yet the next level. What are all the subcomponents that go into making up a good backhand? Foot position, hand placement, racket rotation, etc.

3. Measure the performance of each component

Once you have a list of all the factors that affect elite performance you then need a method of objectively measuring the performance of each subcomponent. Aerobic fitness might be measured using volume of oxygen you can consume while performing at your maximum capacity (VO_2 max), speed might be measured using the 'beep test' and so on.

Now you can measure the effectiveness of any training 'intervention' you think is worthwhile. It often turns out that many 'good ideas' don't work in practice while some unusual ones do.

Using a similar process to improve students' levels of moral development would be a revolution in schools because it would mean the entire school board, advisors and teachers would need to take a much more holistic approach. It would no longer be sufficient to just tinker around the edges with ad hoc discipline or the odd lesson on 'religious studies'.

And while we're at it, we definitely need to reward successful teachers and give them higher status in society than we currently do. It is the old story: you get what you pay for. And if we don't value our teachers as highly as our sports stars or medical doctors then it should come as no surprise to us that the moral health of the next generation of society is going to be poor.

GOVERNMENT POLICY

Governments are notorious for putting in place ad hoc social programmes either to win votes in the next election or in a knee-jerk response to the latest crisis. These programmes often come and go without any lasting impact before they're replaced with another programme that is equally ineffective. Superficial programmes like these must be replaced with a comprehensive approach based on the objective metrics methodology I outlined in the education section above. While this requires far more discipline, effort and planning up front, it's the only technique I've seen that guarantees success.

THE POPULUS AWARD

Finally, I'd like to suggest putting together an independent programme which entitles companies that make a considerable contribution to society the right to display a 'seal' on their products. This idea is similar to that used by the royal family when they issue their 'By Appointment to her Majesty' seal for approved suppliers of products, except in our case the seal is awarded for services to society as a whole. For want of a better title we could call this award the Populus Award. If carefully administered and publicized it could grow to be a valuable award in the same way that awards like the Oscars now exert a powerful influence at the box office. Conversely, it would make sense to award the equivalent of the 'Razzies' for companies that do great harm to society so we can 'name and shame'. After all, the two things every company wants to protect are its bottom line and its public brand image.

PERSONAL ACTIONS

I close this discussion on ethics with a thought my editor came up with after reading this chapter. She found it was an interesting exercise to consider the thousands of decisions and actions she executed each day and to weigh them up on the Moral Scale. Was that last action at Level 3 or Level 4? When I tried this myself I found my results were all over the show. Sometimes I behaved at Level 5 and other times at Level 2. What was interesting to me was that simply being aware of my own levels produced a definite improvement in my daily scores and behaviour. Maybe that elusive goal of 'enlightenment' which everyone talks about could be replaced with something much more tangible and genuinely beneficial.

CHAPTER 20
SOCIETY

Who's in charge of the clattering train?
The axles creak and the couplings strain;
The pace is hot and the points are near,
And sleep has deadened the driver's ear;
And the signals flash through the night in vain,
For Death is in charge of the clattering train.
— Paraphrased from *Punch* and quoted by Sir Winston Churchill

THE GLOBAL FINANCIAL CRISIS
AND HOW IT AFFECTS YOU

I wrote this chapter back in early 2007 at a time when the world's financial markets were powering ahead and Lehman Brothers and Bear Stearns both had AAA ratings. It was a period of euphoria when everyone was full of financial optimism and the American dream powered by capitalism seemed unstoppable. Unfortunately, my friends didn't want me to publish this chapter because my ideas were so at odds with the prevailing mood that they all thought I'd made a mistake.

But as I now tour the suburbs of middle America, I see house after house of good hard-working Americans in foreclosure. Decent citizens' lives have been ruined through no fault of their own.

And yet while they suffer, along with the rest of the world as the global debt crisis mounts, the people who caused this calamity have by and large escaped unharmed. The bankers and brokers still drive around in their fancy cars and get their million-dollar bonuses. Bonuses made possible by the bail outs which were paid from taxes on those who are now homeless.

The problem is, despite all the rhetoric from the bankers in the EU and the politicians in Washington, no one seems to see the true underlying cause. They're all still fixated on the symptoms and don't understand what is fundamentally broken with the model. None of the solutions being proposed is going to do any more than put a small Band-Aid on the gaping wound. If anything, all they have done is kick the problem further down the road and make it harder to solve when the day of reckoning finally comes.

The reason *you* should be so concerned about this is because it affects *you* personally. What happens in the boardrooms of Europe affects the daily lives of everyone all around the world. If a bank defaults in Greece it increases the cost of household mortgages 10,000 kilometres away. A financial policy decision in the Oval Office will ultimately affect a housewife in her kitchen as far away as New Zealand. Fortunately, there are a number of things *you* can do to bring about a permanent solution. You can have your voice heard and you can take back control of your life from those faceless bureaucrats who are continuing to take away your financial security without your permission.

I'll divide this chapter into two main topics:

1. The sharemarket/banking/investment
2. The structure of society.

THE DUTCH TULIP CRASH

Even if you never invest in the sharemarket you'll want to change how it operates because money made and lost there ultimately comes from your pocket — as all those innocent people with houses now in foreclosure can attest to.

Let's start by looking back all the way to the great Dutch Tulip Crash to gain some perspective on recent financial turmoil.

Tulips were first introduced into Holland from Turkey in the sixteenth century when Holland was in its golden age of prosperity, brought about by being the centre of world trade. The novelty of these brightly coloured flowers made them highly sought after by the wealthy as a status symbol. Having tulips in your house was a bit like owning an expensive painting. After a few years of breeding, the Dutch tulips contracted a virus which didn't kill them but caused interesting-coloured patterns to appear on the petals. These patterns came in a wide variety of colours which further increased the rarity of the already unique flowers. Some patterns were thought of as more attractive or rarer than other patterns. Tulips which were already expensive began to steadily rise in value as they became increasingly sought after. Before long people began to speculate on the future price of tulips and traders began to hoard bulbs for the next growing season. Prices were rising so quickly that people began selling their houses and investing their entire life savings into 'tulip futures'. At the height of this frenzy tulips enjoyed a 2000 per cent increase in value in a single month. No wonder everyone was investing in tulips. A worker could sell a single tulip bulb for more than an entire year's wages.

Of course, the prices were not an accurate reflection of the true value of a tulip bulb and so a few wise people decided to sell their tulip investments. This quickly led to a domino effect of falling prices, and before long everyone was trying to sell and no one was trying to buy. The price began to dive, causing people to panic and sell regardless of their losses. The bubble had burst.

Dealers refused to honour contracts and people soon realized they'd traded their homes for a piece of over-hyped vegetation. The Dutch government tried to halt the inevitable crash by guaranteeing to honour 10 per cent of the face value of all contracts. But it was too late. This well-intentioned action by the government of the day, like so many government actions that would follow in times of crisis right up to today, caused the complete meltdown of the entire tulip market. Unfortunately, even those who'd never participated in this crazy market were left penniless as Holland suffered a deep financial depression that would last for generations.

The whole tulip bubble sounds crazy today, but if you think about it, it's no different to the share market crash of 1987, the Dot Com bubble of 2000 or the Housing and Loans credit crunch of 2007.

So what causes the problem and what can *you* do to fix it?

POSITIVE FEEDBACK

The first problem with the sharemarket is that it displays characteristics we can loosely call '*positive feedback*'. Positive feedback occurs when a small movement away from a stable position increases the force, driving you away from the stable position. The easiest way to think of this is to imagine a pencil balancing perfectly on its tip in a stable position. If the pencil now leans a very small amount to the right, gravity will exert a force which pulls it even more to the right. The further it leans over the stronger the force, and so the pencil quickly falls to the ground. This is the opposite of *negative feedback* where the force opposes any movement away from a stable position. The best example of that is a pendulum. The further you pull it away from the bottom the more it wants to return to the bottom.

The problem with positive feedback is that it always causes wild swings because any small change will always get amplified until at some time in the future there is a problem. It's what happens when you get feedback from a microphone when you place it too close to a speaker. A small sound from the speaker is picked up by the microphone which then gets amplified and sent back even louder to the speaker. The noise goes around and around from speaker to microphone until you get a howling screech. A similar thing happens with the sharemarket. If a stock rises people notice it's going up and so they want a part of the action and put in some of their money. But this investment only serves to drive the price up even more, which in turn makes it more attractive for other investors. Before long you've got an upward spiral of positive feedback like the one that drove the tulip craze and the Dot Com bubble. At some point things become so out of whack it's no longer sustainable and the price either stabilizes or falls. As soon as that happens, people want to get out and so the price falls, causing more people to bail and so fuelling the crash.

TIME SCALES

The second problem with the sharemarket, and the reason the brokers and bankers can make money while the average home owner loses their shirt, is because of the time scales involved. Think about it for a moment: what's the difference between investing in the sharemarket and your home mortgage? The answer is: you can buy or sell shares on a minute-by-minute basis but you can't buy and sell your house every few minutes. This discrepancy in time periods between the two systems is why brokers can make money while you can't. As soon as they see a share price rise they buy shares and begin to make a profit on this rising share value regardless of whether the business in which they have bought shares is actually going to make money or not. They only care about the rise of the share price, just as the tulip investors did. The moment they think the shares are going to experience a sustained fall they can bail out and cash in their profits.

But you can't do that with your house. You can't sell your house in New York at 9 a.m. and buy another one at 9.30 and move to Atlanta and then change your mind again at 11.30 and move to Chicago. So the broker can ride the short-term ups and downs and still make a profit while your assets can't be moved anywhere near as quickly. The broker can bail out a day after the stocks begin to fall but the flow through to your house will continue for months and years and there is nothing you can do about it. In fact, brokers *want* the market to have lots of ups and down because they make their money by jumping in and out. The greater the change the more scope there is for profit. Of course, it costs brokers a small fee every time they buy or sell, but it's so small in comparison to the profits they make they can afford to do this on a daily basis if need be. The trouble is, each time they make a profit from speculating they take money out of the system without any tangible benefit. If tulip prices double and then halve and someone wins then other people must lose. It's a zero-sum game and it's the average person who always foots the bill.

THE TIME LIMIT

Now think what would happen if you were not allowed to sell shares in any company for a minimum period of one year after you bought them. Firstly, it would immediately eliminate insider trading! Insider trading occurs when a broker hears about some piece of information regarding a company a few hours or days before everyone else does and then buys or sell shares on the basis of this short-term knowledge advantage. But if shares need to be held for a year at a time, this short-term knowledge becomes virtually meaningless.

The second thing that would happen is that you can't speculate on short-term rises and falls in share prices. You have to think about the *long-term* value of any company you invest in. This would automatically channel money into companies that are going to sustain the economy and provide benefit to society rather than flashy stocks with no real substance. No one would invest in a single tulip bulb at the equivalent of US$100,000 today because you know that price is never going to be sustained a year from now — even if it might reach $120,000 in the next week. Likewise you'd never issue mortgages for four times the value of a house because in a year's time it's likely to be under water. In the same way bonuses on mortgages and financial deals should never be paid at the time of writing the contract but only once the true value of the contract becomes clear a year later. At present brokers and bankers are personally encouraged to write as many contracts as they can because they're paid their bonus immediately. It doesn't matter if the homeowner defaults later on — that's not their problem; they've already got their money and have spent it on holidays, champagne and luxury cars. It's then left to the government and ultimately the tax payer to pick up the pieces.

In mechanical systems which have either positive or negative feedback, the sort of proposal where we delay payments and the ability to sell shares is called 'damping'. It slows down the wild excesses. It's why your car has both springs and shock absorbers. The springs provide the force and the shock absorbers dampen everything down and stop the wild fluctuations. Without shock absorbers you'd end up bouncing down the road after every bump. While there are many ways you can introduce damping into the share market, adding a delay is probably the simplest from a legislative and enforcement point of view and the most powerful. It would work as effectively as the shock absorbers in your car. Then we can all get on with living a comfortable, smooth life and not lurch from crisis to crisis like we seem to be doing every 10 years.

As radical as this suggestion sounds, it could be gradually phased in so that stocks need to be held for one month, then two months, then three months and so on, with a clear timetable for each time increase until the full year is reached. It would transform our entire banking and financial system overnight.

Let's now turn to a much less controversial idea and look at the second area that needs overhauling: the structure of society.

CERTAIN IDEAS SUIT CERTAIN TIMES

When the Wright brothers made the first powered flight their engine could only produce 12 horsepower. This measly power output meant their plane needed to have very light wings with a large surface area in order to provide sufficient lift at the plane's top speed of 30 km/h. Light canvas fabric wrapped over slender beech wood frames provided the perfect solution. But this design and construction would have been totally hopeless if the Wright brothers had strapped a modern

75,000 horsepower F-16 jet engine onto the frame. If they'd done that the wings would have torn clean off the fuselage as soon as they'd opened up the throttle and reached any significant speed. What was perfect for 30 km/h clearly isn't going to work at 300 km/h. And the exact opposite situation is also true. If you put the Wright brothers' engine in a modern aluminium aircraft frame it would never take off because the aluminium fuselage is too heavy and the wings are far too small for the weight and power of the craft. So the wings and the engine need to be perfectly matched. If one develops faster than the other, there's likely to be trouble.

In this chapter we'll see:

- why society *has* been so incredibly successful up until the twenty-first century — the 'wings' and 'engine' were perfectly matched
- why society has recently outgrown the structures holding it together so there is now a real danger the wings will tear off.

Let's start by looking at why we made such spectacular progress during the twentieth century so we can understand why the old system will no longer work in the twenty-first century.

CAPITALISM AND EFFICIENCY

Modern capitalist or free-market societies have evolved and are driven by the relentless pursuit of increased **efficiency**. We all want to get as much as possible for as little as possible. We want the highest-quality products and the latest designs at the cheapest price. Capitalism operates by rewarding those who provide a better service or product at a cheaper price with a steady supply of money. In this scenario efficiency is successful, breeds quickly and drags society happily along with it. Indeed, it is improvements in efficiency which have made Western societies so materially successful compared with other systems.

There are many ways you can improve efficiency but one of the most powerful is to increase **specialization**. Rather than having hundreds of people each grinding their own wheat into flour by hand and baking their own bread, it makes sense to have a single mechanized flour mill doing this on a large scale. Rather than learning how to weld we quickly realize this is a skill that takes years to master but is only infrequently required by most of us. So it becomes much more *efficient* if one person becomes an expert welder while another person becomes a brilliant surgeon. We simply don't have the time or the expertise to master every skill we need, particularly when some are used so infrequently. So the doctor cures hundreds of patients and pays the baker for the bread she eats and pays another 10,000 car assembly workers when she buys herself a new car.

This capitalist drive for efficiency (producing better products at cheaper prices) has been staggeringly successful. You only have to walk into a large American shopping mall and marvel at the millions of quality products you can buy within a few minutes' walk. Computers with 1000 times the power of those used in the first Apollo Moon Program now sell for less than a pair of designer jeans. We're spoilt for choice, quality and price.

Specialization *is* a winner for everyone during the early stages and there is little risk associated with it. But as we become increasingly specialized we also become ever more dependent on a bigger, more complex system over which we have no direct control. The result is that a person living in New York now has absolutely no way of producing their own food, or their electricity, or their fuel, or their water or their clothes. Every resident is totally and utterly dependent on thousands and thousands of people each doing their job correctly and working together as part of some huge unseen team. The bakery can only get its bread to the supermarket if the tanker driver delivers the petrol to the gas station and so on. Initially, this interdependence isn't a problem because there's still a reasonable level of robustness in the system. Instead of everyone baking their own bread we might now have 15 bakeries in the state. If one of these bakeries experiences difficulties this isn't much of a problem because there are still 14 others ready to supply us at a moment's notice. But the nature of a free market is that competition will relentlessly ratchet up the need for ever greater efficiency and this soon leads to **consolidation**. Before long those 15 bakeries have merged to form one or two hugely efficient factories. Again, we're all happy with this state of affairs because we get higher quality bread at a cheaper price.

Consolidation is the natural result of the capitalist drive for efficiency, and it signals that an industry is approaching maturity. If you follow almost any industry you'll find the same story. Whether it's airplane manufacturing or companies that make software operating systems it's only a matter of time before the weaker companies are gobbled up. Instead of the hundreds of car manufacturers and backyard tinkerers that were scattered across America in the 1930s we now have only four or five giants producing high-quality, reliable cars.

Consolidation has another benefit. It allows these huge companies to produce incredibly *sophisticated* products. You only have to lift the hood on a modern car and compare it with one from the 1960s to see how much more high-tech it has become. Simple carburettors that anyone could fiddle with and tune in their home garage are now replaced with complex electronic fuel-injection systems that are controlled by high-powered microprocessors fed by a bewildering array of sensors. It's no longer possible to tune these devices without diagnostic equipment running into the tens of thousands of dollars. Another consequence of this increased sophistication is that it becomes prohibitively expensive for any new company to enter the market. It's no longer possible to start up your own company and expect to sell operating systems in competition with Microsoft because Windows is just too complicated and too mature for any first-time effort to compete against. These stages are shown schematically in the diagram below which I'll call the **complexity curve.**

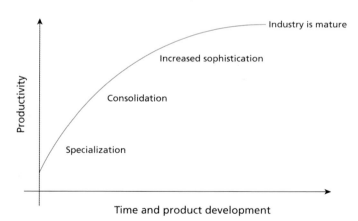

Complexity Curve

MILITARY 'ROBUSTNESS'

Let's now contrast the free-market capitalist system we've just been talking about with the military model. The military operates on a completely different premise to capitalism. Rather than being driven by market forces and the need to increase efficiency, the military's prime requirement is to keep functioning and moving forward, even if large bits of it have been blown to pieces. In short, the military is all about *robustness*. This need for robustness is bred into every nut and bolt of the entire operation. Soldiers are trained to follow orders without thinking because this means that even if the communication lines are destroyed the generals can rely on the last instruction being followed to the letter. That in turn means generals don't have to spend hours guessing what their troops are up to. They know what they're doing and so they can continue planning the battalion's next move with a high degree of confidence. Of course this state of affairs *isn't* efficient because some of the soldiers on the front line might come up with a better plan if they were allowed to think for themselves and make timely decisions based on what's going on in front of them. But if they were allowed to have that sort of autonomy there'd be a risk they'd do something in direct conflict with the general's next plan. The general might bomb his own soldiers because they weren't where he was expecting them to be. So the safe and robust way forward is to 'follow the last order'.

Soldiers are even trained to talk in a robust way at the cost of efficiency. Every sentence is repeated to make sure there are no mistakes:

> Captain: 'Twenty degrees to port and ahead one-third.'
> Engineer: 'Aye, Aye: 20 degrees to port and ahead one-third.'

Repeating every sentence like this certainly isn't an efficient way to speak, but it ensures mistakes like the following humorous Chinese whisper don't occur:

> General from battlefield: 'Send reinforcements, we're going to advance.'
> Received at headquarters: 'Send three and fourpence, we're going to a
> dance.'

The military is also divided up into units, such as divisions or platoons, each with its own complete self-sufficient infrastructure. Again this means there's lots of duplication and lots of redundancy, but it also means it's extremely difficult for the enemy to land one killer blow on a vulnerable weak spot. If one supply route or airport is blown up, the other four supply routes continue unaffected. It's not efficient sending supplies four different ways and supporting four independent managerial teams, but the cost of failure is simply too high for it to be done any other way.

In short, the military builds robustness into its systems, its procedures and of course into its equipment. As a consequence, the military is renowned for being expensive and inefficient — but ultimately that doesn't matter. As long as it keeps on surviving it has a chance of winning the war.

POWERING PAST THE PEAK: NORTHWEST COD FISHING EXAMPLE

Capitalism, with its search for efficiency, *is* good in the initial phase because profit and consumption drives society up the left-hand side of the complexity curve. This situation is what we've all grown up with and it has served us well. It's why in the developed world we live lives of comparative luxury and leisure. But the problem with free markets is that they don't have any sense of when the peak is reached and they keep driving society way past this peak until it becomes brittle

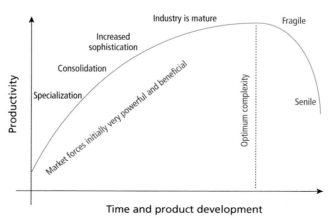

Complexity Curve

and fragile. There are numerous historical examples of this occurring, but somehow we have a knack of brushing them aside as exceptions and don't take their significance on board. Here's one example.

The Northwest Atlantic Ocean was one of the richest fishing grounds in the world. For over 100 years fishermen landed an average of 250,000 tons of fish a year. But from the 1950s onward, market forces fuelled a steady increase in fishing rates until the annual catch reached around 800,000 tons in 1968. The market forces driving these increases included the opening up of foreign markets, which had an insatiable appetite for the cod, and the development of bigger trawlers that dragged nets bigger than football fields to catch the cod. Demand drove supply and led to increasing efficiency, sophistication and market consolidation. But the waters of the Northwest couldn't sustain those forces because the fish couldn't breed fast enough and so by the 1990s the entire marine ecology of the Northwest Atlantic had collapsed. By 1994 the total biomass of all fish left swimming in the entire Northwest that year was now only 1700 tons! The same sea that had sustainably supported an annual catch of 250,000 tons for more than 100 years now had only 1700 tons of fish in it.

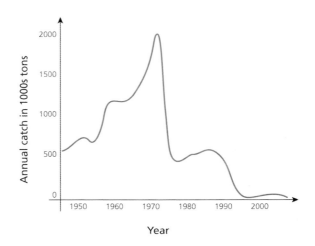

This is not an unusual story. Countless similar examples have been documented of industries or societies booming right up to the very last minute before they collapsed. The American grasslands supported millions of bison before they were farmed so extensively that the grasslands collapsed into arid dust bowls that haven't recovered even to this day. This illustrates two vital points:

- Market forces are able to generate tremendous growth and prosperity in the initial cycle but they are *blind* to long-term robustness.
- Just because something has always worked doesn't mean it will continue to work.

INCREASED DEPENDENCE AND FRAGILITY

The second problem caused by increased specialization, consolidation and complexity is that society becomes less and less robust. This was graphically brought home to me when I was living in England in 2000 and a small group of people blockaded Britain's oil refineries in protest over increased fuel prices. Within days, 90 per cent of Britain's petrol stations were empty and by the end of the week hospitals had run out of fuel for their ambulances, forcing emergency calls to go unanswered. There wasn't even enough fuel for the back-up generators to cover intensive care operations and so many were cancelled. Supermarkets quickly ran out of food and within nine days England was brought to a virtual standstill, all because one small part of the infrastructure had been disrupted.

This shows the problem of going too far in terms of specialization and consolidation — of going over the top of the efficiency curve. While a small number of massive refineries is more efficient than many small refineries, and while delivering fuel according to a 'Just in Time' model is more cost-effective than having substantial reserves on hand, the downside is that it provides minimal robustness in case of some disruption.

Another example of this increasing fragility in modern society was the power outage that swept across America in 2006. Millions of people were plunged into darkness and society stopped because a *single* relay switch failed in *one* city. When that city called on its neighbouring city for back-up generation the second city didn't have enough reserve either, and so it also became overloaded. In quick succession, city after city fell like a stack of dominos because not a single city had sufficient reserve to halt the collapse. This was a clear demonstration that most American cities had pushed too far up the curve of cost-efficiency at the expense of robustness.

Another problem with increased complexity and specialization is that everything depends on hundreds of components all working perfectly together and within very small margins of error. If any one of them fails, the whole system grinds to a halt.

Other signs the infrastructure isn't sufficiently robust are when the roads are jammed solid during the daily commute or when water is rationed in a country like England despite its long rainy winters. Sure, none of these problems is easy to deal with. Fixing leaking pipes and building more reservoirs is incredibly expensive and takes dozens of years before they are completed. Similarly, public transport requires enormous vision if it is to meet not only the needs of the current generation but also of the *next* generation who will only start using it long after the politician has retired. Finding solutions today is never going to be as simple as it was in the old days when you could just put down a few more lanes in the open countryside. Our cities are now packed solid and there's no more space to build additional roads.

It is probably fair to say that most Western societies are now operating with less reserve and less robustness than at any time since World War II. This means any sudden increase in stress could easily cause a catastrophic failure on a scale not seen before. While it's not the scope of this book to cover new stresses in detail it's probably worthwhile mentioning a few in passing.

THE ENERGY BOTTLENECK

Historically, there have been various periods named for their typifying technology, such as the Stone Age and the Bronze Age. No matter how old you are, you've always lived in the 'Oil Age' where almost everything you do is powered by cheap energy. Unfortunately, like all the 'Ages' before this one, the Oil Age as we know it will soon begin to decline. By this I don't mean we will 'run out' of oil. There will be still be oil, but our society will no longer be powered by and depend upon oil in the same way that it does today. And that's going to create a huge strain on society.

The reason why oil has helped us obtain such a high standard of living is because of the huge amount of energy each cupful of petrol contains. A gallon of gasoline contains enough energy to push a 1 ton car 50 km. That's a colossal amount of energy. Try pushing a car on your own for that distance and you'll soon get the idea of the amount of energy petrol contains. In fact a single tank of petrol is able to do as much physical work as a fit labourer working 12 hours a day, every day of the year, for four years! That's part of the reason why we enjoy such a high standard of living today. This huge amount of power in petrol also explains why electric cars are so difficult to make. A tank full of petrol is equivalent to 120 fully charged car batteries costing around $6000 and which can only be charged a limited number of times before they wear out.

Another way to look at the power of petrol is to consider the amount of corn you need to produce enough biofuel to fill your car's petrol tank just once. A single tank of gas would require a farm big enough that it could feed a person for an entire year. Think about it — each tankful of gas is equivalent to an entire year's worth of groceries.

We harness this energy from petrol in bulldozers and cranes to build our houses and roads and in manufacturing to produce all the modern luxuries of life we take for granted. It's been a key part of what is called the Industrial Revolution. At present this energy is easily obtained. We simply dig a hole in the ground and we pump the oil out. It's like we've inherited a bank vault full of money buried in the ground. We haven't had to work for this money and we can just make withdrawals to fund our extravagant lifestyle whenever we want. The problem is we're not putting anything back into that vault and so one day our funds will begin to run low.

A little thought shows that almost every aspect of modern society is underpinned by cheap liquid energy. Even the price of something as simple as a McDonald's cheeseburger depends on cheap oil. The meat patties come from cows which grazed on fields that were ploughed and sowed by tractors burning diesel. More diesel is then used to transport the cows from the farm to a central abattoir miles away where they are killed before the meat is sent to yet another place

for processing into patties. There is a similar story for all the other ingredients such as the wheat in the bun and the lettuce and the pickles. Likewise, every worker in the entire food chain, from those ploughing the field and rounding up the cows to those serving you at the McDonald's outlet, also relies on fuel to get to work each day and home again at night.

By the time you put a cheeseburger in your mouth almost a third of the cost of the burger is in the fuel needed to make the ingredients and get it to you. If the cost of fuel goes up, sooner or later this increase must be reflected in the cost of your cheeseburger.

Some people argue we'll find more oil, but since the 1980s we've found very little new high-quality crude that is easy and cheap to recover. In fact oil companies have recently downgraded their 'proven reserves' because they'd historically exaggerated them in an effort to increase the value of their company's share price. Most experts now agree we've reached the peak in the amount of oil we can take out of the ground each day. We hit this limit because the rate at which the oil flows towards the drill hole suddenly slows once an underground oil well falls below a certain level. Twenty years ago there were 15 oilfields able to supply 1 million barrels of oil a day. Now there are only four left. The unfortunate thing is that we've hit this limit at exactly the same time as the economies of China and India are booming, producing a huge spike in demand. There is now simply no slack left in the system and this brings with it another problem. Oil is primarily concentrated in geopolitically unstable regions and the distribution of this oil runs through a number of critical 'choke points'.

Half of the world's oil is distributed by tankers and most of them travel through five narrow and vulnerable straits. Some 15 million barrels of oil pass through the Strait of Malacca each day and another 16 million barrels go through the Strait of Hormuz. If any one of these 'choke points' experiences difficulties the entire world would experience an oil meltdown. Western civilization simply cannot function without a huge daily supply of oil. And that of course makes these choke points vulnerable to political interference or terrorist attacks.

The lifeblood of our society, oil, therefore suffers simultaneously from two critical problems:

1. Insufficient supply
2. Lack of robustness in delivery.

POPULATION DEMOGRAPHICS

Another major stress about to hit Western societies is an unprecedented shift in population demographics. What is important here is the ratio of people aged between 20 and 65 who are working compared with those who are not working but who still consume resources.

$$\text{Ratio} = \frac{\text{Number of people not working (pensioners and children)}}{\text{Number of workers}}$$

As a society changes from one that is rapidly expanding to one that has a stable population, there are inevitably fewer people born as mothers have fewer children and have them later in their life. The simplest way to visualize how this is changing is to consider a pyramid which shows how many children under 20 there are compared with how many retired pensioners there. The diagram below shows how this is expected to change for a country like Japan in the next 15 years compared with what it used to look like in 1950.

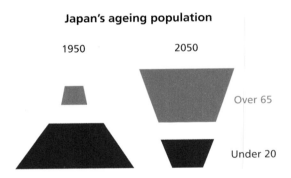

Japan's ageing population

1950   2050

Over 65

Under 20

In 1950 there were approximately 9.3 children under the age of 20 for every adult aged over 65. In 2050 this number is expected to fall to half a child for every pensioner. With fewer children growing up, the future work force is reducing while at the same time life expectancy for pensioners is increasing.

So we have more people consuming with less people producing. What compounds this issue is that not only are there fewer workers and more pensioners but each individual pensioner is now consuming vastly more resources than before because they are living longer. Instead of living for 5–10 years after retirement pensioners are now living for a further 20–30 years, owing to advances in medical care.

But the medical care they are receiving is becoming increasingly complex and expensive, which places yet another burden on already overstretched resources. The net result is that governments around the world must allocate an increasingly larger proportion of their budget to pensions, leaving less for all the usual things such as schools, roads, healthcare, policing and so on — the very things that guarantee society's robustness.

THE GOLDEN AGE

Once we understand these curves one question we must ask ourselves is: *Where are we on the complexity curve?* Are we still climbing up the left-hand side of the graph or are we already tumbling down the right-hand side? The following observations give us some clues as to where we are on the curve. We can no longer afford to run the supersonic Concorde and we can't afford to send people to the moon any more. Life expectancy, which was continually increasing, has reached a limit and is now starting to shorten again in many countries. We're spending a larger percentage of our money on basics such as food, housing and petrol. While we're more technically advanced than we've ever been, the *quality* of our life is no longer increasing. If anything it's becoming more of a struggle.

None of these observations provides us with definitive answers as to where we are on the efficiency curve, but they do sound some warning bells and tell us we should look more closely at our society and not just sail on as we always have. Maybe it's time to sacrifice some level of efficiency for increased robustness in our financial and political systems and to look at the longer term when making decisions that affect our cultural heritage. Maybe our fixation on using GDP as a measure of our progress is no longer appropriate and instead we need to institute accurate measures of society's robustness and resilience to various challenges such as foreign debt, oil supply, geopolitical stability, and import/export restrictions, to name but a few.

> '*It was the best of times, it was the worst of times, ...*' — Charles Dickens, *A Tale of Two Cities*

A NEW TYPE OF ACCOUNTING

The following hypothetical story shows why we need to embrace a new way of thinking about society. Imagine a team of builders has just spent six months building, fitting out and painting a lovely new house. Then as soon as it's finished another team of workers comes along and pulls it down and grinds the whole house into fine dust, which is then buried. Clearly, a tremendous amount of energy, labour and skill will have been expended in this fruitless exercise. Washing machines, light bulbs and furniture will have been designed, manufactured and shipped, not to mention all the building materials and construction required to erect the house itself. But by the time everything has been ground back to dust nothing of any value to society will have actually been produced despite all this effort. Even though invoices and paper money flowed back and forth between contractors and subcontractors, even though lawyers and accountants were employed to keep track of everything and despite many natural resources being consumed — the net result was *nothing*. There was a great *illusion* of employment and progress but society as a whole gained absolutely no benefit. Yet, according to the standard accounting measures most governments use, this pointless exercise helped increase the gross domestic product of that country and contributed to employment figures.

Clearly, this is an extreme example but it raises the important question of the *long-term value to society* of everything we do. We need to think about the *social utility* or value to society of commerce and not simply the dollar value we attach to everything. Such a realignment would automatically assign a greater value to prevention than it would to cure. Initiatives to cut smoking, binge drinking and to increase exercise would be given priority over costly operations to repair perfectly predictable lifestyle diseases that could have been prevented in the first place. Building gleaming new hospitals and stocking them with expensive medical equipment, scanners and highly skilled doctors whose job it is to treat patients' diseases brought on by preventable lifestyle abuses is as effective as building houses and tearing them down again.

The challenge with instituting this new type of social accounting is that it requires more intelligence to operate because we don't just blindly add up the dollars as they circulate through society any more. We need to know how *effective* those dollars were in the *long run* compared with all other options. It's a whole new way of thinking for our politicians. We no longer think about how many patients our health budget treated in a year or how long the waiting lists were, but rather how many patients we *didn't* need to treat in the first place because they were now healthier as a result of better social policies. We no longer think in terms of how many new cars were imported and sold, how many billions of dollars of oil used and how many miles of new roads were built (all huge contributors to standard measures of prosperity), but how easily and comfortably each person can get to wherever it is they want to get whenever they want to. After all, going where you want when you want is the aim of transport. Having a $150,000 car doing 10 miles per gallon while stuck in a traffic jam for an hour shows we've actually gone backwards compared with 10 years ago if the same journey then used to take only 20 minutes. All those new cars, roads and fuel have added nothing.

We need to assign a weighting value that scales every dollar of public money spent so it reflects the true long-term value of enhancing the quality of life for each individual in society. Using that sort of accounting system shows that while developing a network of electric-powered monorails and underground trains has a high initial cost it becomes the least expensive option after sufficient time because it reduces:

- the number of foreign cars and spare parts imported to keep each of them going
- the billions of gallons of foreign oil imported and burnt
- the millions of tons of carbon dioxide produced by all that fuel, which has an ongoing carbon, environmental and health cost
- the ongoing maintenance of all the roads
- the number of car crash victims who need hospital treatment.

If at the same time a monorail provides society with a better quality of life and faster transport to and from our cities, it's clearly a winner compared with buying more cars.

SUB-OPTIMAL SOLUTIONS

Market forces aren't only blind to robustness and knowing when the peak is reached; they're also blind as to what is ultimately *good* for us. A simple example would be the decision to import cheap products manufactured overseas. In the short term we may get equal-quality products at a cheaper price, but in the long term we may end up destroying our own manufacturing base, causing widespread unemployment and creating huge foreign debt. Then, when our country is unable to compete, the foreign country can increase the price and we're in no position to respond. In this case market forces have driven us entirely in the wrong direction because they have no sense of what is *ultimately* good for us. The only thing driving market forces are 'which decisions today make the graph go higher tomorrow'. We can represent this by imagining a graph with two options shown in red and blue that separate at a fork or choice. Clearly, the red option produces the greatest immediate benefit, but if we extend this graph further into the future it may turn out that the red curve isn't the best option after all.

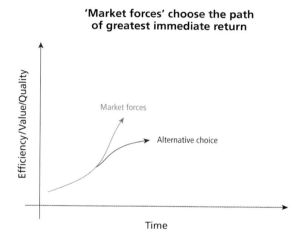

**'Market forces' choose the path
of greatest immediate return**

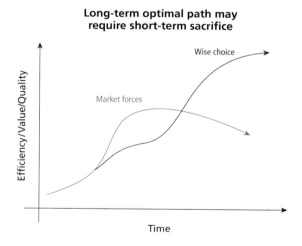

**Long-term optimal path may
require short-term sacrifice**

It's only by rising above simplistic predictions based on short-term trends and by having a deep insight into the complexities of our existence that we are able to choose the optimal path. This is not something that can be left to blind 'market forces'.

CHAPTER 21
A PERSONAL CHALLENGE TO YOU

I hope this book stimulated you to think in new ways and challenged you to re-evaluate some of your beliefs. If so, I have achieved half of what I set out to do.

What is even more important to me is that you take action and stand up for Truth. To speak out whenever Truth is violated or false beliefs are allowed to fester unchallenged. For you to speak out on behalf of society so that we leave a better place for our successors than we inherited. Not just from a material point of view but from one which is measured according to the metrics of beauty, peace, happiness and knowledge.

This is an intensely satisfying and honourable journey.

I invite you to join me.

Kerry